FRACTIONAL DISTILLATION

Laboratory Scale Chemistry through Pilot Plant Operations

by

Sydney Young, D.Sc., F.R.S.

Professor of Chemistry
University College, Brighton

Wexford Press
2008

PREFACE

DURING the past eighteen years I have been engaged in investigations which necessitated the preparation of chemical materials in the purest possible state, and as the great majority of these substances were liquids, the process of fractional distillation had, in most cases, to be resorted to for their purification.

The difficulties I met with in some of the separations led me to make a careful investigation of the whole subject, and I was thus enabled to devise some new methods and forms of apparatus which have been described from time to time in various scientific journals.

It is in the hope that the solution of the difficulties which so often occur in carrying out a fractional distillation may be rendered easier, and that the value and economy of highly efficient still-heads in laboratory work may come to be more widely recognised than is generally the case at present, that this book has been written.

My sincere thanks are due to Professor J. Campbell Brown for the loan of valuable ancient works by Libavius and Ulstadius, from which Figures 2 and 34 have been taken; to my colleague, Dr. F. E. Francis, for reading the proofs, and for much valuable assistance in compiling the index ; and to Professor R. A. Gregory and Mr. A. T. Simmons, B.Sc., for many useful suggestions regarding the arrangement of the MS. and for the perusal of the proofs.

In the description and illustration of the stills employed in commerce I have derived much assistance from articles in Thorpe's " Dictionary of Applied Chemistry" and Payen's " Précis de Chimie Industrielle."

I have made much use of the experimental data of Brown, Konowaloff, Lehfeldt, Zawidski and other observers, and have, as far as possible, made due acknowledgment, but in some of the tables this has not been practicable.

Several fractional distillations and numerous experiments have been carried out while the book was being written and the results have in many cases not been published elsewhere.

S. Y.

CONTENTS

CONTENTS

LIST OF ILLUSTRATIONS

FRACTIONAL DISTILLATION

CHAPTER I

INTRODUCTION

Object of Distillation.—The object of distillation is the separation of a volatile liquid from a non-volatile substance or, more frequently, the separation of two or more liquids of different volatility.

If only one component of a mixture is volatile, there is no difficulty in obtaining it in a pure state by distillation, and in many cases the constituents of a mixture of two or more volatile liquids may be separated—though frequently at much cost of time and material—by means of the simple apparatus described in this chapter. For the fractional distillation of such complex mixtures as petroleum or fusel oil, the improved still-heads described in Chapters X to XII must be employed.

Quantitative Analysis by Distillation.—The determination, by ordinary analytical methods, of the relative quantities of two or more organic compounds in a mixture is often a matter of great difficulty, but, in many cases, the composition of the mixture may be ascertained approximately and, not seldom, with considerable accuracy from the results of a single distillation, if a very efficient still-head be

B

employed.   It is hoped that this method may prove of considerable value.

Difficulties Encountered.

Difficulties Encountered.—The subject of fractional distillation is full of interest owing to the fact that difficulties so frequently occur, not only in the experimental work, but also in interpreting the results obtained.

In the distillation of petroleum, such difficulties are of common occurrence and are due to one or other of three causes :—(a) to the presence of two substances, the boiling points of which are very close together ; (b) to the presence of one or more components in relatively very small quantity ; (c) to the formation of mixtures of constant boiling point.

The separation of two liquids which boil at temperatures even 20° or 30° apart, such as ethyl alcohol and water, or benzene and isobutyl alcohol, may be impossible owing to the formation of a mixture of minimum or, less frequently, of maximum boiling point.   It is, indeed, only in the case of substances which are chemically closely related to each other that the statement can be definitely made that the difficulty of separating the components of a mixture diminishes as the difference between their boiling points increases.

In any other case, we must consider the relation between the boiling points, or the vapour pressures, of mixtures of the substances and their composition, and unless something is known of the form of the curve representing one or other of these relations, it is impossible to predict whether the separation will be an easy one or, indeed, whether it will be possible.

The form of these curves depends largely on the chemical relationship of the components, and it is now possible, in a moderate number of cases, to form an estimate, from the chemical constitution of the substances, of the extent to which the curves would deviate from the normal form, and therefore to predict the behaviour of a mixture on distillation.

Fractional distillation is frequently a very tedious process and there is necessarily considerable loss of material by evaporation and by repeated transference from the receivers to the still, but a great amount of both time and material may be saved by the use of a very efficient still-head ; and when the object of the distillation is to ascertain the composition of a mixture, very much greater accuracy is thereby attained.

APPARATUS.

Ancient Apparatus.—The process of distillation is evidently a very ancient one, for Aristotle (1)* mentions that pure water may be obtained from sea-water by evaporation, but he does not explain how the condensation of the vapour can be effected. A primitive method of condensation is described by Dioscorides and by Pliny, who state that an

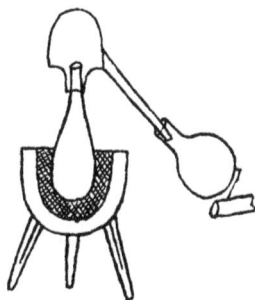

FIG. 1.—Alexandrian still with head, or *alembic*.

FIG. 2.—Ancient still with water condenser.

oil may be obtained by heating rosin in a vessel, in the upper part of which is placed some wool. The oil condenses in the wool and can be squeezed out of it.

The Alexandrian chemists added a second vessel, the head or cover, called by the Arabians the *alembic*, to the boiler

* The numbers in brackets refer to the bibliography at the end of the chapters.

or still, and a simple form of apparatus used by them is shown in Fig. 1.

Later on, the side tube was cooled by passing it through a vessel containing water. The diagram, Fig. 2, is taken from Libavius, *Syntagma Alchymiæ Arcanorum*, 1611.

Modern Apparatus.—The apparatus employed at the present time is similar in principle, but, in addition, a thermometer is used to register the temperature. In Fig. 3

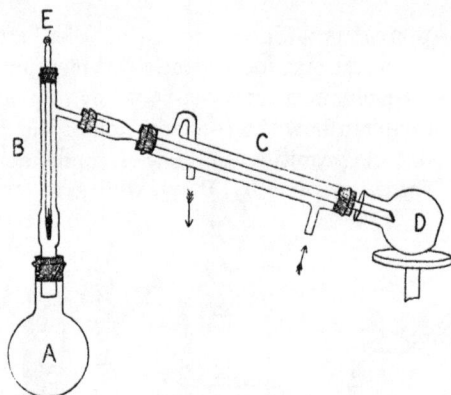

Fig. 3.—Ordinary still, with Liebig's condenser.

the ordinary form of apparatus is shown, and we may distinguish the following parts :—The still, A; the still-head, B; the Liebig's condenser, C, in which the vapour is deprived of heat by a current of cold water; the receiver, D; the thermometer, E. The still is usually heated by means of a Bunsen burner.

The flask or still is fitted with a cork through which passes the still-head, and the side delivery tube from the still-head passes through a second cork in the condensing tube. For liquids which boil at a high temperature, or which act chemically on cork, it is more convenient to have the still and still-head in one piece and to elongate the delivery tube so that it may pass, if necessary, through the Liebig's condenser (Fig. 4).

The Still.—If a glass flask is used it should be globular, because a flat-bottomed flask is liable to crack when heated with a naked flame. It should not be larger than is necessary for the amount of liquid to be distilled.

The Still-head.— The still-head should not be very narrow, or the thermometer may be cooled slightly below the temperature of the vapour. It is a good plan to seal a short length of wider tubing to the still-head near the bottom, leaving a sufficient length of the narrower tubing below to pass through the cork in the still, as shown in Fig. 3.

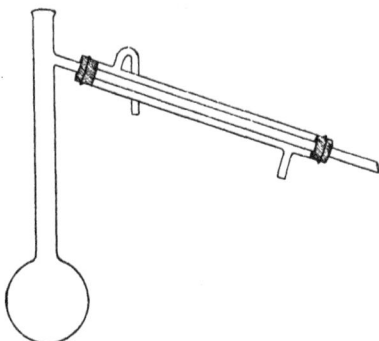

FIG. 4.—Modified form of still with condenser.

The still-head, as supplied by dealers, is often too short. It should, if possible, be long enough for the thermometer to be placed in such a position that not only the mercury in the bulb but also that in the stem is heated by the vapour of the boiling liquid ; otherwise a troublesome and somewhat uncertain correction must be applied (p. 12), and, if the distillation is not proceeding quite steadily, a little air may be carried back from time to time as far as the thermometer bulb and the temperature registered by the thermometer will then fluctuate and will, on the whole, be too low (p. 28).

The Condenser.—If the boiling point of the liquid to be distilled is higher than about 170°, the condensing tube should not be cooled by running water for fear of fracture. A long tube should be used and the cooling effect of the surrounding air will then be sufficient.

When a Liebig's condenser is used there is no advantage in having either the inner or the outer tube very wide ; an internal diameter of 7 or 8 mm. is sufficient for the inner, and of 15 mm. for the outer tube. If the outer tube is much wider it is unwieldy, and, when filled with water, it is inconveniently heavy. A mistake that is rather frequently made may be referred to here. It is usual to seal a short wide tube to the long, narrow condensing tube for the insertion of the delivery tube from the still-head. The tubes are often sealed together in such a way that when the distillation is proceeding a little pool of liquid collects at the junction (Fig. 5 a), and, in the frac-

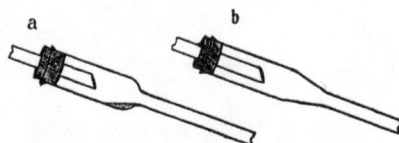

FIG. 5.—Condensing tube of (a) faulty, (b) correct, construction.

tional distillation of a small quantity of liquid, the error thus introduced may be serious. The fault is easily remedied by heating the wide tube close to the junction with the narrow one until the glass is soft and then drawing it out very gently until it has the form shown in Fig. 5 b.

When a long still-head is used, it is advisable to bend the narrow tube just below its junction with the wider one, so that the condenser may be vertical in position instead of sloping gently downwards. Much less space is thus taken up on the laboratory bench, and the receivers are somewhat more conveniently manipulated.

The Source of Heat.—For laboratory purposes an ordinary Bunsen burner is usually employed. Wire gauze, asbestos cardboard, sand baths, or water or oil baths are not, as a rule, to be recommended, because the supply of heat can be much more easily regulated without them, and a round-bottomed flask, if properly blown, is so thin walled that there is no danger of fracture when the naked flame is applied. The flask should be so placed that the flame actually comes in contact with the bottom of it; this is especially necessary when the liquid to be distilled is liable to "bump." Many substances, such as carbon disulphide, which boil quite regularly under the ordinary atmospheric pressure, bump more or less violently when the pressure is greatly reduced, unless special precautions are taken. Under a pressure of 361 mm. carbon disulphide boils at 25°, and if a quantity of it be distilled under this pressure with the flame placed some distance below the bottom of the flask, it may happen that the whole of the carbon disulphide will pass over without any ebullition whatever taking place. The liquid, however, in these circumstances, becomes considerably superheated, and if a bubble does form there will be a sudden and perhaps violent rush of the extremely inflammable vapour. If, however, the top of the burner be placed only about 2 mm. below the bottom of the flask, so that the minute flame touches the glass, ebullition will take place quietly and regularly.

There are liquids which cannot be prevented from bumping in this way, and the best plan is then to add a few small fragments of porous porcelain [a clay pipe broken in small pieces answers the purpose very well] or pumice-stone, or both, or a number of small tetrahedra of silver or platinum. A water or oil bath need only be used when a solid substance is present in the flask as, for instance, when a liquid is distilled over lime or phosphorus pentoxide, or when the liquid is liable to decompose when heated with the naked flame.

It is customary to employ a water-bath for the distillation of ether, but it is doubtful whether this is necessary or even advisable except in the case of the ethereal solution of a solid substance or one that will not bear heating much above 100°. When an accident occurs it is almost invariably because, owing to " bumping," or to the distillation being carried on too rapidly, some of the vapour escapes condensation and comes in contact with a flame in the neighbourhood, generally that below the water-bath. If a naked flame were used the distillation could be much more easily regulated, and there would probably be really less danger than if a water-bath were employed.

Protection of Flame from Draughts.—In order that satisfactory results may be obtained it is necessary that

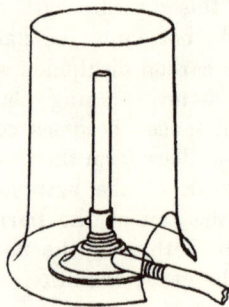

Fig. 6.—Simple flame protector.　　　Fig. 7.—Still with steam jacket.

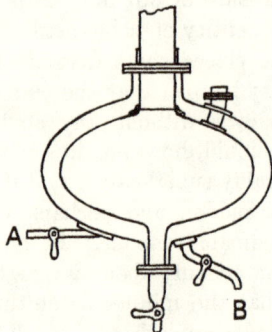

the distillation should proceed with great regularity and the heat supply must therefore not be subject to fluctuations. The most important point is to guard against draughts, and, to do this, the ordinary conical flame protector may be used, or a simple and efficient guard may be made from a large beaker by cutting off the bottom and taking a piece out of the side (Fig. 6).

Steam as Source of Heat.—On the large scale, the still is frequently heated by steam under ordinary or increased pressure (Fig. 7). The steam is introduced through the pipe A, and the condensed water is run off at B.

The Thermometer.—In carrying out a fractional distillation one must be able, not only to read a constant or nearly constant temperature with great accuracy, but also to take readings of rapidly rising temperatures. These requirements are best fulfilled by the ordinary mercurial thermometer, which is therefore, notwithstanding its many drawbacks, used in preference to the air or the platinum resistance thermometer. If accurate results are to be obtained the following points must be attended to.

1. **Calibration.**—The thermometer must be carefully calibrated, and it would be a great advantage if all thermometers were compared with an air thermometer, for two mercurial thermometers, constructed of different varieties of glass, even if correct at 0° and 100°, will give different and incorrect readings at other temperatures, more especially at high ones, for various reasons:

(*a*) In the first place, it is impossible to obtain an absolutely cylindrical capillary tube, and therefore the volume corresponding to a scale division cannot be quite the same in all parts of the tube. Various methods have been devised for calibrating the stem (2, 3), but even when this is done there remain other sources of error.

(*b*) The position of the mercury in the stem at any temperature depends on the expansion both of the mercury and the glass, and, for both substances, the rate of expansion increases with rise of temperature.

(*c*) Different kinds of glass have different rates of expansion, so that two thermometers made of different materials —even if the capillary tubes were perfectly cylindrical— would give different readings at the same temperature. It is therefore necessary to compare the readings of a mercurial thermometer with those of an air thermometer, or of another mercurial thermometer which has previously been standardised by means of an air thermometer. Or, instead of this, a number of fixed points may be determined by heating the thermometer with the vapours of a series of pure liquids boiling under known pressures.

Table 1 contains a list of suitable substances with their boiling points, and the variation of temperature for a difference of 10 mm. from the normal atmospheric pressure.

TABLE 1.

Substance.	Boiling-point under normal pressure.	Variation of temperature per 10 mm. pressure.
Carbon disulphide	46·2°	0·40°
Ethyl alcohol	78·3	0·33
Water	100·0	0·37
Chlorobenzene	131·95	0·50
Bromobenzene	156·15	0·51
Aniline	184·4	0·51
Naphthalene	218·05	0·58
Quinoline	237·45	0·59
Bromonaphthalene	280·45	0·64
Benzophenone	305·8	0·63
Mercury	356·75	0·75
Sulphur	444·55	0·87

In this way a table, or curve, of corrections may be constructed, and the error at any scale reading of the thermometer may be easily ascertained.

"Normal" thermometers may now be purchased; they are compared with a standard thermometer before graduation, and true temperatures are said to be registered by them.

2. **Redetermination of Zero Point.**—The zero point of a thermometer should be redetermined from time to time, as it is subject to changes which, in the case of the cheap soda glass thermometers, may be considerable. These changes are of two kinds :

(a) If a thermometer be graduated shortly after the bulb has been blown, the zero point will be found to rise, at first with comparative rapidity, then more and more slowly, and the elevation of the zero point may go on for many years. If the thermometer be kept at a high temperature—especially, as shown by Marchis, if there are periodical, slight fluctuations of temperature—the rise of the zero point takes place with much greater rapidity, and up to, at any rate, 360°, and probably 450°, the higher the temperature the more rapid is the rise and, apparently, the higher is the final point reached. A rise of more than 20° has several times been observed in the case of soft German glass thermometers on being subjected to prolonged heating at 360°. In all cases the rise, which is rapid at first, becomes slower and slower, and it seems doubtful whether, at any given temperature, actual constancy of zero point has ever yet been attained. If, however, a thermometer has been heated for many hours to a given high temperature and then allowed to cool very slowly, subsequent heating to lower temperatures has very little effect on the zero point. The best thermometers, as first recommended by Crafts, are kept at a high temperature for a long time before being graduated.

(b) If a thermometer—even after its zero point has been rendered as constant as possible—be heated and then cooled very rapidly, a slight fall of zero point will be observed; but after a day or two the greater part of this fall will be recovered, and the remainder after a long period.

3. Volatilisation of Mercury in Stem of Thermometer.—
In the cheaper thermometers there is a vacuum above the mercury and, when the mercury in the stem is strongly heated, volatilisation takes place, the vapour condensing in the cold, upper part of the tube; when, therefore, the temperature is really constant it appears to be gradually falling. The better thermometers, which are graduated up to high temperatures, contain nitrogen over the mercury, a bulb being blown near the top of the capillary tube to prevent too great a rise of pressure by the compression of the gas; but thermometers which are only required for moderate temperatures, say, not higher than 100° or even 150°, are not usually filled with nitrogen. If, however, such thermo-

meters are used for the distillation of liquids boiling at so low a temperature as 100°, or even 80°, a quite perceptible amount of mercury may volatilise and, after prolonged heating, errors amounting to 0·2° or 0·3° may occur. It would be much better if all thermometers required to register temperatures higher than 60° were filled with nitrogen.

4. **Correction for Unheated Column of Mercury.**—As already mentioned, the thermometer should, if possible, be so placed in the apparatus that not only the mercury in the bulb but also that in the stem is heated by the vapour of the boiling liquid; otherwise the following correction, which, at the best, is somewhat uncertain, must be applied:—

To the temperature read, add $0·000143(T-t)N$, where T is the observed boiling point, t the temperature of the stem above the vapour, and N the length of the mercury column not heated by the vapour, expressed in scale divisions.

The coefficient 0·00016—the difference between the cubical expansion of mercury and that of glass—is very frequently employed, but it is found in practice to be too high; and Thorpe has shown that the value 0·000143 gives better results.

Table 2, on page 13, given by Thorpe (4) may be found useful.

5. **Superheating of Vapour.**—When the amount of liquid in the still is very small, the vapour is liable to be superheated by the flame, and unless the bulb of the thermometer is thoroughly moistened with condensed liquid, too high a temperature will be registered. If a very little cotton wool, or, for temperatures above 230°, a little fibrous asbestos, be wrapped round the bulb of the thermometer, it remains, as a rule, thoroughly moist, and, with a pure liquid, heated by a naked flame, the thermometer registers a perfectly con-

TABLE 2.

$T-t$	\(N\) 10	20	30	40	50	60	70	80	90	100	110	120	130	140	150	160	170	180	190	200
10	0·01	0·03	0·04	0·06	0·07	0·09	0·10	0·11	0·13	0·14	0·16	0·17	0·19	0·20	0·21	0·22	0·24	0·26	0·27	0·29
20	0·02	0·06	0·09	0·11	0·14	0·17	0·20	0·22	0·26	0·29	0·31	0·34	0·37	0·40	0·43	0·46	0·49	0·51	0·54	0·57
30	0·04	0·09	0·13	0·17	0·21	0·26	0·30	0·34	0·39	0·43	0·47	0·51	0·56	0·60	0·64	0·68	0·73	0·77	0·82	0·86
40	0·05	0·11	0·17	0·23	0·28	0·34	0·40	0·47	0·52	0·57	0·63	0·69	0·74	0·80	0·86	0·91	0·97	1·03	1·09	1·14
50	0·07	0·14	0·21	0·29	0·36	0·43	0·50	0·60	0·64	0·71	0·79	0·86	0·93	1·00	1·07	1·14	1·22	1·29	1·36	1·43
60	0·08	0·17	0·25	0·35	0·43	0·51	0·60	0·70	0·77	0·86	0·94	1·03	1·12	1·20	1·29	1·37	1·46	1·54	1·63	1·72
70	0·10	0·20	0·30	0·40	0·50	0·60	0·70	0·80	0·90	1·00	1·10	1·20	1·30	1·40	1·50	1·60	1·70	1·80	1·90	2·00
80	0·11	0·22	0·34	0·45	0·57	0·68	0·80	0·91	1·03	1·14	1·26	1·37	1·49	1·60	1·72	1·83	1·94	2·05	2·17	2·29
90	0·13	0·26	0·39	0·51	0·64	0·77	0·90	1·03	1·16	1·30	1·42	1·54	1·66	1·80	1·93	2·05	2·17	2·31	2·45	2·54
100	0·14	0·28	0·43	0·57	0·71	0·85	1·00	1·14	1·29	1·43	1·58	1·71	1·84	2·00	2·15	2·29	2·43	2·57	2·72	2·86
110	0·16	0·31	0·47	0·63	0·79	0·94	1·10	1·26	1·42	1·58	1·73	1·89	2·04	2·20	2·36	2·51	2·67	2·83	2·99	3·15
120	0·17	0·34	0·51	0·69	0·86	1·03	1·20	1·37	1·54	1·71	1·89	2·06	2·23	2·40	2·57	2·74	2·92	3·09	3·26	3·43

stant temperature until the last trace of liquid in the bulb has disappeared.

With a water or oil bath the danger of superheating is greater, and the cotton wool may become dry at the end of the distillation. In that case, the temperature registered may be too high, though, as a rule, the error is not so great as it would be if the bulb were not protected.

. 6. **Correction of Boiling Point for Pressure.** —The barometer must always be read and corrected to 0° (p. 269) and, in a long distillation or in unsettled weather, it may be necessary to read it frequently, for the boiling point of a liquid varies greatly with the pressure.

It is impossible to give any accurate and generally applicable formula for correcting the observed boiling point to that under normal pressure (760 mm.), but the following may be taken as approximately correct :—

$$\theta = 0{\cdot}00012(760 - p)(273 + t)$$

where θ is the correction in centigrade degrees to be added to the observed boiling point, t, and p is the barometric pressure (5).

This correction is applicable, without much error, to the majority of liquids, but for water and the alcohols a better result is given by the formula

$$\theta = 0{\cdot}00010(760 - p)(273 + t)$$

Crafts (6) has collected together the data for a number of substances, from which the values of c in the formula $\theta = c(760 - p)(273 + t)$ are easily obtained.* Table 3 (p. 15) contains the boiling points on the absolute scale, T, the values of c and those of dp/dt for some of the substances referred to by Crafts and also for a considerable number of additional ones (7).

---

* There are a few misprints in the table given by Crafts and, since it was published, many additional accurate determinations of boiling point and vapour pressure have been made.

The values marked with an asterisk have been determined indirectly, and are not to be regarded as so well established as the others.

TABLE 3.

Substance.	T.	dp/dt	c	Substance.	T.	dp/dt	c
Oxygen	90·3°	75·9*	0·000146*	Iodobenzene	461·45	18·0	0·000120
Nitrogen	77·5	89·0*	0·000145*	Bromonaphthalene	553·45	15·75	0·000115
Argon	86·9	83·2*	0·000138*	Methyl ether	249·4	32·0	0·000125
Chlorine	239·4	33·2	0·000126	Ethyl ether	307·6	26·9	0·000121
Bromine	331·75	25·2*	0·000120	Acetone	380·0	26·4	0·000115
Iodine	458·3	18·75	0·000116	Benzophenone	578·8	15·8	0·000109
Mercury	629·8	13·4	0·000118	Anthraquinone	650·0	13·6	0·000113
Sulphur	721·4	12·2	0·000114	Aniline	457·4	19·6	0·000112
Ammonia	240·1	37·7	0·000110	Quinoline	510·5	17·0	0·000115
Sulphur dioxide	262·9	33·7	0·000113	Methyl formate	304·9	28·8	0 000114
Carbon disulphide	319·25	24·7	0·000127	Ethyl formate	327·3	26·6	0·000115
Boron trichloride	291·25	26·8	0·000128	Propyl formate	353·9	24·5	0·000115
Phosphorus trichloride	346·85	23·45	0·000123	Methyl acetate	330·1	26·8	0·000113
Carbon tetrachloride	349·75	23·25	0·000123	Ethyl acetate	350·15	25·1	0·000114
Silicon tetrachloride	329·9	24·0	0·000126	Propyl acetate	374·55	23·5	0·000114
Stannic chloride	387·1	21·4	0·000121	Isobutyl acetate	389·2	22·5	0·000114
Methane	109·0	68·2*	0·000135	Methyl propionate	352·7	24·9	0·000114
n-Pentane	309·3	25·8	0·000125*	Ethyl propionate	372·0	23·7	0·000113
n-Hexane	341·95	28·9	0·000122	Propyl propionate	395·15	22·3	0·000114
n-Heptane	371·4	22·8	0·000121	Isobutyl propionate	409·8°	21·4	0·000114
n-Octane	398·8	21·1	0·000119	Amyl propionate	433·2	20·4	0·000113
Isopentane	300·95	26·2	0·000127	Methyl butyrate	375·75	23·3	0·000114
Di-isobutyl	382·1	20·9	0·000125	Ethyl butyrate	392·9	22·3	0·000114
Hexamethylene	353·9	22·7	0·000124	Propyl butyrate	415·7	20·9	0·000115
Benzene	353·2	23·45	0·000121	Amyl butyrate	451·6	19·5	0·000113
Toluene	383·7	21·75	0·000120	Methyl isobutyrate	365·3	23·8	0·000115
Ethyl benzene	409·15	20·8	0·000120	Isobutyl isobutyrate	419·6	20·6	0·000116
Naphthalene	491·0	17·1	0·000119	Methyl alcohol	337·7	29·6	0·000100
Anthracene	616·0	15·0	0·000108	Ethyl alcohol	351·3	30·35	0·000094
m-Xylene	412·0	21·1	0·000115	Propyl alcohol	370·4	28·8	0·000094
Triphenyl methane	626·0	14·8	0·000108	Amyl alcohol	403·0	25·3	0·000098
Methyl chloride	249·35	31·9	0·000126	Phenol	456·0	20·5	0·000107
Ethylene dibromide	405·0	20·8	0·000119	Acetic acid	391·5	23·9	0·000107
Fluorobenzene	358·2	23·3	0·000120	Phthalic anhydride	559·0	16·0	0·000112
Chlorobenzene	405·0	20·5	0·000120	Water	373·0	27·2	0·000099
Bromobenzene	429·0	19·3	0·000120				

When a boiling point is to be corrected, the constant c for the substance may usually be found by reference to Table 3. Either the constant for that substance in the table most closely related to the one under examination is to be used, or the constant may be altered in conformity with one or other of the following generalisations.

Relation of Constant c to Molecular Weight and Constitution.—1. In any homologous series, or any series of closely related substances, except the alcohols, acids, phenols, the lower esters and perhaps some others, the higher the molecular weight the lower is the constant. Examples :— The normal paraffins ; methyl and ethyl ethers ; toluene and meta-xylene ; acetone and benzophenone ; benzene, naphthalene and anthracene.

2. Iso-compounds have higher values than their normal isomerides, and if there are two iso-groups the value is still higher. Examples :—Isopentane and normal pentane; di-isobutyl and n-octane ; methyl isobutyrate and methyl butyrate.

3. When hydrogen is replaced by a halogen, the value is lowered. Examples :—Benzene and a mono-derivative ; naphthalene and bromonaphthalene.

4. By replacing one halogen by another no change is usually produced. Examples :—The four halogen derivatives of benzene.

5. All compounds containing a hydroxyl group—alcohols, phenols, water, acids, have very low values. But the influence of the hydroxyl group in lowering the constant seems to diminish as the complexity of the rest of the molecule increases. Thus, with methyl, ethyl, and propyl alcohols the constants must be as much as $0 \cdot 000035$ lower than those of the corresponding hydrocarbons, but with amyl alcohol it is only $0 \cdot 000029$, and with phenol only $0 \cdot 000015$ lower. On the other hand, the constant tends in general to be lowered as the molecular complexity increases, and these two factors, acting in opposite directions, neutralise each other more or less completely ; thus, in the case of the alcohols at any rate, there is apparently no relation between the values of the constant and the molecular weight.

6. The esters—formed from alcohols and acids—have rather low values and here again the constant is nearly independent of the molecular weight.

Modifications of the Still.—For ordinary laboratory purposes a round bottomed glass flask is the most convenient form of still, but if a large quantity of liquid has to be distilled, especially when it is very inflammable, it is safer to employ a metal vessel. Metal vessels are generally made use of on the large scale.

Any alteration in the shape of the still is merely a matter of convenience and does not call for special mention.

Modifications of the still-head are of great importance and will be considered later in Chapters X to XIII.

Modifications of the Condenser.—For liquids boiling above the ordinary temperature, but below about 170°, the straight Liebig's condenser is employed. If a liquid boils at a higher temperature than 170°, there would be danger of fracture if the glass delivery tube were cooled by water. The cooling effect of the surrounding air is, however, sufficient if a long tube be employed. For very volatile liquids, the delivery-tube must be cooled by ice or by a freezing-mixture (pounded ice and salt, or ice and concentrated hydrochloric acid are convenient for moderately low temperatures). In this case a spiral, or " worm," tube should be used (Fig. 8).

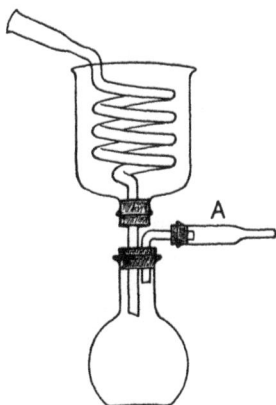

Fɪɢ. 8.—Condenser for volatile liquids.

Condensation of moisture in the receiver is prevented by the drying tube, ᴀ.

Modifications of the Receiver.—If a liquid boils at a very high temperature, or if it suffers decomposition at its ordinary boiling point, it may be necessary to distil it under

reduced pressure. For cases of simple distillation the
apparatus shown in Fig. 9 may be employed, but if the

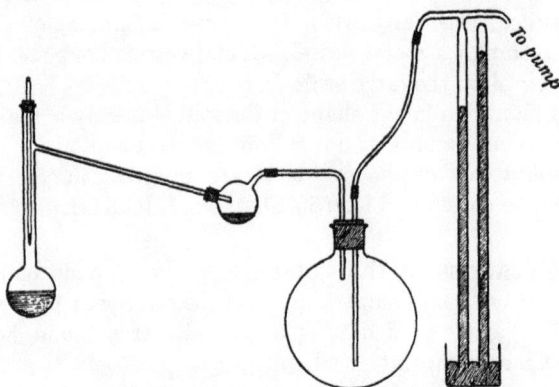

FIG. 9.—Simple apparatus for distillation under reduced pressure.

distillate is to be collected in separate portions, the removal
of the receiver would necessitate admission of air into the
apparatus and a fresh exhaustion after each change. [The
large globe in Fig. 9 serves to keep the pressure steady and
to prevent oscillation of the mercury in the gauge.] Various
methods have been devised to allow of the receivers being
changed without altering the pressure, of which the following
may be mentioned.

1. **Thorne's Apparatus.**—A series of stopcocks may be
arranged in such a manner that air may be admitted into
the receiver and a fresh one put in its place while the
distillation bulb remains exhausted (Fig. 10). The stopcock *b* is closed, and *c* is turned so as to admit air into the
receiver, which is then disconnected and a fresh one is put
in its place. The stopcock *a* is then closed to shut off the
still from the pump, and *c* is turned so as to connect the
pump with the new receiver, which is then exhausted until

the pressure falls to the required amount, when a and b are again opened.

This method, though comparatively simple, is attended by several disadvantages; there is some risk of leakage when so many stopcocks are used— even when a three-way stop-cock is employed (as in Fig. 10) in place of two simple ones— and this is especially the case because ordinary lubricants cannot as a rule be used for b, through which the condensed liquid flows. Moreover, the changing of the receiver, the manipulation of the stopcocks, and the exhaustion of the fresh

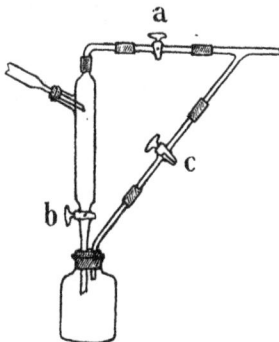

Fig. 10.—Thorne's apparatus for distillation under reduced pressure.

receiver take up some time, during which the progress of the distillation cannot be closely watched.

It is better to employ some method by which the change of receiver may be effected without admitting air at all, and there are two forms of apparatus in general use by means of which this may be done.

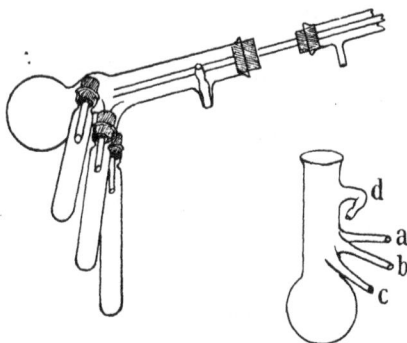

Fig. 11.—Bredt's apparatus for distillation under reduced pressure.

2. **Bredt's Apparatus.**—To the end of the delivery tube from the still-head is attached a round-bottomed flask with a long neck to which are sealed three narrow tubes a, b, and c, approximately at right angles to it (Fig. 11), and a fourth

tube d, which serves to admit air when the distillation is completed. The bulb of the flask serves as one receiver, and each of the three narrow tubes is connected with a cylindrical vessel by means of a perforated cork. The long-necked flask is first placed with the three receivers in an inverted position, so that the first fraction collects in the flask ; when a change is to be made, the neck of the flask is rotated until the drops of distillate fall into one of the cylindrical receivers and each of these in turn can be brought vertically below the end of the delivery tube.

3. **Brühl's Apparatus.**—A number of test-tubes are placed in a circular stand which may be rotated within an

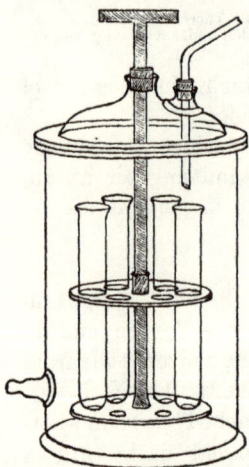

exhausted vessel (Fig. 12) so that any one of the tubes may easily be brought under the end of the delivery tube. This arrangement seems on the whole to be the most convenient, as the change of receiver can be effected with the greatest ease and rapidity. Many other forms of apparatus have been devised, but they do not differ in principle from one or other of the three described above.

Prevention of Leakage.— When a liquid is distilled under reduced pressure, it is necessary that all joints in the apparatus should be air-tight. India-rubber stoppers cannot well be used for

FIG. 12.—Brühl's apparatus for distillation under reduced pressure.

the still because this substance is attacked or dissolved by so many organic liquids, and ordinary corks are seldom quite air-tight. Page (8), however, finds that all leakage may be effectually prevented by first exhausting the apparatus and then covering the cork with the ordinary liquid gum, sold in bottles (not gum arabic, which is apt to crack when dry). The gum may be conveniently applied with a brush and, if necessary, the application may be repeated several times.

REFERENCES

1. Kopp, *Geschichte der Chemie*, Beitrage I., 217.
2. "Methods employed in Calibration of Mercurial Thermometers,"
   British Association Report for 1882, 145.
3. Guillaume, "Traité pratique de la Thermométrie de précision,'
   p. 112.
4. Thorpe, "On the Relation between the Molecular Weights of Substances and their Specific Gravities when in the liquid state," *Trans. Chem. Soc.*, 1880, **37**, 159.
5. Ramsay and Young, "Some Thermodynamical Relations," *Phil. Mag.*, 1885, [V.], **20**, 515.
6. Crafts, "On the Correction of the Boiling Point for Barometric Variations," *Berl. Berichte*, 1887, **20**, 709.
7. Young, "Correction of the Boiling Points of Liquids from observed to normal Pressure," *Trans. Chem. Soc.*, 1902, **81**, 777.
8. Page, "Cork *versus* Rubber," *Chem. News*, 1902, **86**, 162.

CHAPTER II

THE BOILING POINT OF A PURE LIQUID

THE STATICAL METHOD

THERE are two methods by which the "boiling point" of a liquid under a given pressure may be determined, the *statical* and the *dynamical.* By the first method the pressures exerted by the vapour of the liquid at a series of temperatures are ascertained and plotted against the temperatures, a curve, the vapour pressure curve, being then drawn through the points (Fig. 13).

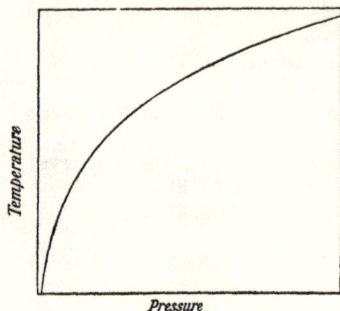

FIG. 13.—Vapour pressure curve.

This curve has a two-fold meaning; it represents not only the vapour pressures of the liquid at different temperatures, but also the boiling points of the liquid under different pressures. Thus, the vapour pressure of water at 50° is 91·98 mm., and water boils at 50° under a pressure of 91·98 mm.

Evaporation in Absence of Air.—The vapour pressures of a liquid at temperatures lower than its boiling point under atmospheric pressure may be determined by placing some of the liquid over the mercury in a barometer tube and heating the tube to different temperatures. The difference between the height of the barometer and that of the column of mercury in the tube, after correcting for the expansion of the mercury, and, if necessary, for its vapour pressure, gives the vapour pressure of the liquid.

It is necessary to take great care that the liquid introduced is quite free from dissolved air, otherwise, under the reduced pressure and at the higher temperature, some of this air would be expelled and the measured pressure would be the sum of the pressures of the vapour and of the air. If the liquid is pure and free from air, the pressure varies only with the temperature and does not depend on the relative volumes of liquid and vapour so long as both are present. This constancy of vapour pressure is, in fact, a very delicate test of the purity of a liquid.*

Evaporation in Presence of Air.—On the other hand, Regnault has shown that almost exactly the same amount of evaporation will take place into a vacuous space as into the same space containing any gas which does not act chemically on the liquid and is only very slightly soluble in it. The only difference is that evaporation into a vacuum usually takes place almost instantaneously, whereas in presence of a gas the process is a slow one, owing to the time required for diffusion between vapour and gas.

If, then, the barometer tube contained air, and we were able to measure its pressure, we could calculate the true pressure of the vapour by subtracting that of the air from the total pressure.

Dalton's law of partial pressures is, in fact, applicable to this case, and in constructing the vapour pressure curve

---

* As will be seen later, there are certain liquid mixtures which behave in many respects like pure liquids ; at a given temperature the vapour pressure of such a mixture would not depend on the relative volumes of liquid and vapour, but at a different temperature the vapour pressure would no longer be quite independent of the volumes.

from such results, it is the partial pressures of the vapour, and not the total pressures, that must be plotted against the temperatures.

THE DYNAMICAL METHOD

A liquid is said to boil when it is in a state of ebullition, but the scientific term "boiling point" is not applied to the temperature of a boiling liquid. The temperature of ebullition depends partly on the gaseous, or other, pressure on its surface, partly on the vertical depth of the liquid, and partly on the cohesive force between its molecules and the adhesion of the liquid to the walls of the vessel; it would be very difficult, if not impossible, to frame a convenient definition to include all these factors.

Ebullition.—If we heat a liquid in an ordinary glass flask by means of a Bunsen burner placed below it, the formation of bubbles of vapour at the lower surface of the liquid, in contact with the glass where the heat is received, is facilitated by the presence of air dissolved in the liquid or adhering as a film to the glass and by sharp points, or roughnesses, on the surface of the glass.

If a minute bubble of air is formed, it will serve as a nucleus for a larger bubble of vapour, but in order that the bubble may increase in size by evaporation from the liquid surrounding it, it is clear that the vapour must overcome the pressure of the column of liquid above it as well as that of the atmosphere. Now the pressure exerted by the vapour of a liquid, in contact with that liquid, depends solely on the temperature, and therefore, under the most favourable conditions, the temperature of the liquid surrounding the bubble must be so high that the vapour pressure is equal to the sum of the pressures of the atmosphere and of the column of liquid.

Cause of "Bumping."—But if the liquid is very free from air, and if the walls of the vessel are very smooth and

clean, bubbles are formed with much greater difficulty and
the temperature of the liquid may rise much higher; it is
then said to be *superheated*, and when a bubble forms, the
vapour pressure corresponding to the temperature of the
liquid is much greater than the sum of the pressures of the
atmosphere and of the column of liquid; consequently,
vapour is evolved and the bubble increases in size with
great rapidity, and at the same time the temperature of the
liquid falls to some extent. Under these conditions the
liquid boils irregularly and is said to "bump."

The liability to intermittent ebullition is still greater when
the liquid is covered by a layer of another lighter but less
volatile liquid, as when water has a layer of oil over it: if a
globule of well-boiled water is immersed in a bath of oil of the
same specific gravity, so that the water is in contact only
with the oil, the formation of bubbles is very difficult
indeed, and the temperature may be raised far above
100°; when a bubble is at last formed, the whole of the
water may be suddenly converted into steam with explosive
violence.

Definition of the term "Boiling Point."—By the
term *boiling point* is to be understood the highest temperature attainable by a liquid, under a given pressure of its
own vapour, when evaporating with a perfectly free surface,
and when the heat reaches the surface from without. Thus, if
we cover the bulb of a thermometer with cotton wool or other
porous material, saturate the wool with the liquid under
examination and suspend the thermometer in a test tube
heated in a bath to a temperature at least 20° higher than
the boiling point of the liquid, the temperature will rise,
vaporisation will take place, and the air in the test tube will
be expelled by the vapour. Under these conditions, so long
as the cotton wool remains thoroughly moistened by the
liquid, the temperature cannot rise above a maximum which
is not influenced by that of the bath, but depends solely on

the pressure of the vapour and therefore, since these are equal, on the atmospheric pressure.

This maximum temperature is the true "boiling point" of the liquid under a pressure of its vapour equal to that of the atmosphere. It will be observed that the boiling point of a liquid can only be correctly determined by observing the temperature of the liquid itself under such conditions that ebullition is impossible.

Determination of Boiling Point.—The true boiling point of a liquid is identical with the condensing point of its vapour under the same pressure, provided that some liquid is present and that the vapour is not mixed with an indifferent gas or vapour, and it is usually more convenient to measure the condensing point of the vapour than the boiling point of the liquid. To do this an ordinary distillation bulb is generally employed (Fig. 3.) The walls of the vertical tube give up heat to the surrounding air, and some of the vapour condenses, the remainder being therefore in contact with condensed liquid. Condensation also takes place to a slight extent on the thermometer, unless the amount of liquid in the still is very small, when the vapour is liable to be superheated, owing to the flame or the heated gases from it, playing against the dry walls of the vessel, and as the thermometer loses heat only very slowly by radiation it may, under these conditions, become too hot for vapour to condense on it. The vapour is especially liable to be superheated when a water or oil bath is employed as the source of heat.

Condensation on the thermometer is promoted by covering it with a little cotton wool (1), or, for high temperatures, fibrous asbestos and, in determining a boiling point or carrying out a distillation with the ordinary apparatus, it is always advisable to cover the thermometer bulb in this way. If the liquid, the boiling point of which is to be determined, is not known to be quite pure, it is advisable to distil a fair

quantity of it ; if it is pure the boiling point should remain
quite constant during the whole distillation ; if not, the
temperature will rise (unless we are dealing with a mixture
of constant boiling point), and some idea of the nature and
amount of the impurity will be gained by observing the
extent of the rise and whether it takes place in the early or
late stages of the distillation, or both, or whether there is a
steady rise during the whole period.

Effect of Impurities on Boiling Point.—If there is
a rapid rise at first and the temperature afterwards remains
steady or nearly so, it may be assumed that there is a much
more volatile liquid present. If the temperature is steady at
first but rises rapidly when the distillation is nearly complete, the conclusion may be drawn that a much less volatile
liquid is present. In either case the steady temperature will
approximate very closely to the true boiling point, but it
would be more satisfactory, especially in the second case, to
collect the best portion of the distillate separately, and to
redistil it and read the boiling point again. If, on the other
hand, there is a fairly steady rise of temperature throughout,
the presence of one or more substances, not very different
in volatility from the pure liquid itself, is probable, and it
is impossible to ascertain the boiling point of the pure liquid
without carrying out a fractional distillation to remove the
impurities.

Reduction of Boiling Point to Normal Pressure.—
Assuming that the liquid is pure, that the thermometer has
been compared with an air thermometer, and that the
precautions mentioned in the last chapter have been
attended to, the corrected temperature will give the true
boiling point of the liquid under a pressure equal to that
of the atmosphere at the time. It is frequently necessary,
however, to compare the boiling point of the liquid with
that observed by another experimenter, or to compare it

with that of some other liquid, and it is therefore convenient
to ascertain what the boiling point would be under normal
pressure and, in stating the result for future reference, to
give this reduced boiling point.

The method of correcting the boiling point of a liquid from
observed to normal pressure is given on p. 14.

Prevention of "Bumping."—It frequently happens,
especially when the distillation has to be carried out under
greatly reduced pressure, that the liquid is liable to boil
with bumping (p. 24) and two methods have been proposed
to prevent this.

1. As already mentioned, small tetrahedra of silver or
platinum, or fragments of porous material such as unglazed
porcelain or pumice-stone, may be added to facilitate the
formation of bubbles, or

2. A slow current of air may be admitted through a
capillary tube which passes nearly to the bottom of the
vessel.

The latter method is an excellent one for preventing the
bumping, but it must be remembered that an error is caused
by the presence of the air which is introduced. The true
boiling point of the liquid and the condensing point of the
vapour depend on the pressure of the vapour itself and not
necessarily on the total gaseous pressure to which the liquid
is exposed. If there is air mixed with the vapour, the total
pressure remains unaltered but the partial pressure of the
vapour is diminished, and the observed temperature is lower
than the boiling point under the read pressure. That this
is so may be easily proved by altering the rate at which air
enters through the capillary tube; the more rapid the in-
troduction of air the lower will be the observed temperature,
and unless the amount of air is exceedingly small the
observed boiling point will be sensibly too low.

Law of Partial Pressures.—Dalton's law of partial
pressures is, in fact, applicable to the boiling point of a

liquid as, indeed, is evident from the fact that in determining a boiling point by the statical method, if air is mixed
with the vapour, it is the partial pressure of the vapour and
not the total pressure that must be taken into account
(p. 23). Many experiments might also be described to
prove the truth of this statement, but it will be sufficient to
mention the following :—

Experimental Proofs.—Water, when distilled in the ordinary manner under a pressure of 15 mm., boils at about
18° ; but Schrötter (2) observed so long ago as 1853 that
when some water was placed in a shallow clock glass, sup-
ported on a short tripod on a second clock glass containing
strong sulphuric acid, the whole being placed under a bell-
jar which could be exhausted by an air-pump, the temperature fell to − 3° when the pressure was reduced to 15 mm.
Here the aqueous vapour was rapidly absorbed by the strong
sulphuric acid, so that, when the total pressure was 15 mm.,
the partial pressure of the vapour in contact with the water
must have been only about 4 mm. The " boiling point " of
the water therefore fell below the freezing point ; but in
these circumstances, bubbles could not be formed (except
possibly air bubbles), for the vapour pressure of the water
would be far lower than the total gaseous pressure.

More rapid diffusion and removal of vapour, as well as
freer evaporation, was effected by suspending in the bell-jar
a thermometer, the bulb of which was covered with a piece
of sponge soaked in water ; in this case, under a pressure of
40 mm., at which the boiling point of water under ordinary
conditions is 34°, the temperature actually fell to − 10°.
Here the partial pressure of the vapour must have been only
a very small fraction of the total pressure.

The following experiment (3) affords a still more striking
proof of the correctness of the statement that the " boiling
point " does not necessarily depend on the atmospheric
pressure. A copper air bath, through which a current of

air could pass freely, was heated to 205°; and a thermometer,
the bulb of which was covered with cotton-wool moistened
with boiling water, was suspended in the bath through an
opening in the top. The pressure of the atmosphere was
748 mm., and the water on the cotton-wool was in a strongly
heated chamber, yet its temperature, instead of remaining
at nearly 100°, fell rapidly to 66°, and remained nearly constant at this point. When, however, the current of air
through the bath was retarded by closing the grating in the
side, the temperature of the water rose to about 80°, and
when steam was introduced into the bath so as to replace
the air as completely as possible by aqueous vapour, the
temperature of the water rose to 99°, though that of the
bath had fallen slightly. Lastly, on allowing some of the
steam to escape, the temperature of the water fell again
to 80°.

In this experiment, it is clear that the temperature of the
water did not depend on that of the air bath or on the
atmospheric pressure, both of which remained nearly constant. By limiting the supply of air and by introducing
steam, nothing was altered but the relative pressures of the
aqueous vapour and of the air surrounding the water. The
greater the partial pressure of the vapour, the higher was
the temperature reached by the water, and when the air
was almost completely replaced by aqueous vapour, the
temperature rose very nearly to the ordinary boiling point
of water.

Spheroidal State.—It is well known that when a drop
of water is allowed to fall on a red-hot sheet of metal, such
as platinum, it does not touch the metal, nor does it boil
although rapid evaporation takes place, but it assumes the
spheroidal state, moving about over the hot surface like a
globule of mercury on a table. Careful experiments (4) have
shown that the temperature of the water under these conditions does not reach 100°, and the explanation of this fact

is probably that the aqueous vapour surrounding the spheroidal drop is diluted with air, so that its partial pressure is less than the total atmospheric pressure. There are two reasons, either of which would be sufficient, why ebullition cannot take place :—

1. Heat is received from without towards the surface, from the whole of which evaporation can take place freely, as there is no contact between the water and the heated platinum ;

2. The vapour pressure corresponding to the temperature of the water is lower than the atmospheric pressure, so that a bubble, if formed, would at once collapse.

Wet and Dry Bulb Hygrometer.—Again, the difference in temperature which is usually observed between the wet and dry bulb in the ordinary hygrometer depends on the difference between the partial pressure of the aqueous vapour in the air and the maximum pressure possible at the temperature of the dry bulb, that is to say, the vapour pressure of water at that temperature.

Non-miscible Liquids.—Lastly, as will be shown later on, the fact that when two non-miscible liquids are distilled together, the boiling point is lower than of either component when distilled alone, may similarly be explained by the law of partial pressures, the vapour of each liquid acting like an indifferent gas towards the other.

REFERENCES

1. Ramsay and Young, "On a New Method of Determining the Vapour Pressures of Solids and Liquids," *Trans. Chem. Soc.*, 1885, **47**, 42.

2. Schrötter, "On the Freezing of Water in Rarefied Air," *Liebig's Annalen*, 1853, **88**, 188.

3. Young, "Sublimation," *Thorpe's Dictionary of Applied Chemistry*, III., 611.

4. Balfour Stewart's *Treatise on Heat*, 6th Edition, 129.

CHAPTER III

Influence of Molecular Attractions on Miscibility of Liquids and on Heat and Volume Changes during Admixture.—In studying the behaviour of two liquids, A and B, when mixed together, one should consider

1. The attraction of the like molecules—of those of A for each other and of those of B for each other,

2. The mutual attraction of the molecules of A and B.

If the attraction of the unlike molecules is relatively so slight as to be negligible, one may expect the liquids to be non-miscible or very nearly so. With a somewhat greater relative attraction between the unlike molecules there would be miscibility within small limits, and it is reasonable to assume that in the process of mixing there might be slight absorption of heat and slight expansion.

Generally, in the comparison of various pairs of liquids, as the mutual attraction of the unlike increases relatively to that of the like molecules, one would expect increasing and finally infinite miscibility; absorption of heat at first, diminishing to zero and changing to increasing heat evolution; and diminishing expansion followed by increasing contraction. These various changes do not, in many cases, run strictly *pari passu*, and, among liquids which are miscible in all proportions, it is not unusual to find a small amount of contraction attended by slight heat absorption, as for example, when a little water is added to normal propyl alcohol; but in the case of certain closely related chemical compounds, such as chlorobenzene and bromobenzene, there is neither any appreciable change of volume nor any measurable evolution or absorption of heat when the liquids are mixed together. For such substances it is probable that the different molecular attractions, A for A, B for B, and A for B, are very nearly equal, and that the relation suggested by D. Berthelot (1) and by Galitzine (2), namely that $a_{1.2} = \sqrt{a_1.a_2}$, holds good. [$a_{1.2}$ represents the attraction of the unlike molecules and a_1 and a_2 the respective attractions of the like molecules.]

There would appear then, to be two simple cases :—

1. That in which the attraction represented by $a_{1.2}$ is relatively so slight that the liquids are practically non-miscible ;

2. That of two closely-related and infinitely miscible liquids which show no heat or volume change when mixed together.

The vapour pressures of some mixed liquids and of non-miscible pairs of liquids have been determined by Regnault, Magnus, Konowaloff and other observers.

Non-miscible Liquids

It was shown by Regnault that when two non-miscible liquids are placed together over the mercury in a barometer tube, the observed vapour pressure is equal to the sum of those of the two liquids when heated separately to the same temperature. Each liquid, in fact, behaves quite independently of the other and, so long as both are present in fair quantity and one is not covered by too deep a layer of the other, it does not matter what are their relative amounts or what are the relative volumes of liquid and vapour. If, however, the upper layer is deep, the maximum pressure may not be reached for a considerable time unless the heavier liquid by shaking or stirring is brought to the surface to facilitate its evaporation.

Partially Miscible Liquids

In the case of two partially miscible liquids the vapour pressure was found to be less than the sum of those of the components, but greater than that of either one singly at the same temperature.

INFINITELY MISCIBLE LIQUIDS

The vapour pressures of many pairs of infinitely miscible liquids have been determined by several experimenters and, as with the changes of volume and of temperature on mixing the liquids, so with the vapour pressures of the mixtures, very different results are obtained in different cases. There can be no doubt that the behaviour of mixtures, as regards vapour pressure, depends on the relative attraction of the like and the unlike molecules. When the mutual attraction of the unlike molecules is not much more than sufficient to cause infinite miscibility—for example, with normal propyl alcohol and water—the vapour pressure, like that of a partially miscible pair of liquids, may be greater than that of either component at the same temperature. On the other hand, when that attraction is relatively very great (formic acid and water) the vapour pressure of the mixture may be less than that of either component. It seems reasonable to suppose that, when the attractions of the like and unlike molecules are equal or nearly so, the relation between vapour pressure and composition should be a simple one, and the question what is the normal behaviour of mixtures has been discussed by several investigators.

Normal Behaviour of Mixtures. Guthrie.—Guthrie (3) concluded that if we could find two liquids showing no contraction, expansion, or heat change on mixing, the vapour pressures should be expressed by a formula which may be written $P = \dfrac{wP_A + (100 - w)P_B}{100}$ where P, P_A and P_B are the vapour pressures of the mixture, and of the two components A and B, respectively, at the same temperature, and w is the percentage *by weight* of the liquid A. In other words, the relation between the vapour pressure and the percentage composition by weight should be represented by a straight line.

Speyers.—Speyers (4) concludes that the relation between vapour pressure and *molecular* percentage composition is always represented by a straight line when the molecular weight of each substance is normal in both the liquid and gaseous states. The equation, $P = \dfrac{m\,P_{A} + (100 - m)\,P_{B}}{100}$ where m is the molecular percentage of A, should then hold good.

Van der Waals.—Van der Waals (5) considers that the last-named relation is true when the critical pressures of the two liquids are equal and the molecular attractions agree with the formula proposed by Galitzine and by D. Berthelot, $a_{1.2} = \sqrt{a_{1}.a_{2}}$.

Guthrie was clearly in error in taking percentages by weight, and Speyer has certainly made his statement too general, for there are cases known in which the relation does not hold, although both the liquids have normal molecular weights (for example, n-hexane and benzene, or carbon tetrachloride and benzene).

Kohnstamm's Experiments.—In order to test the correctness of the conclusion arrived at by Van der Waals, Kohnstamm (6) has determined the vapour pressures of various mixtures of carbon tetrachloride and chlorobenzene, the critical pressures of which, 34,180 mm. and 33,910 mm., are nearly equal, and he finds that the curvature is not very marked.

At the temperature of experiment, the maximum deviation from the straight line amounted to about 6 mm. on a total calculated pressure of 76 mm., or about 7·9 per cent. It is probable that, in this case, the formula $a_{1.2} = \sqrt{a_{1}.a_{2}}$ does not accurately represent the facts.

Closely Related Compounds.—From a study of the alteration of volume produced by mixing various pairs of liquids together, F. D. Brown (7) concluded that this change would probably be smallest in the case of closely related chemical compounds and he obtained some indirect evidence in favour of this view; it seems a fair assumption that for such mixtures this, and other physical relations, should be of a simple character.

Experimental Determinations by Statical Method.—Direct measurements of vapour pressures by the statical method have, however, in nearly every case been carried out with mixtures of liquids which have no very close chemical relationship, but Guthrie (*loc. cit.*) made such determinations with mixtures of ethyl bromide and ethyl iodide, and his results make it probable, though not certain, that the equation $P = \dfrac{mP_A + (100 - m)P_B}{100}$ holds good for this pair of substances. It is not known, however, whether the critical pressures of ethyl bromide and ethyl iodide are equal, though it is very probable that they may be.

Linebarger (8) has determined the vapour pressures of a few mixtures of each of the following pairs of liquids by drawing a current of air through them (p. 80):—benzene and chlorobenzene, benzene and bromobenzene, toluene and chlorobenzene, toluene and bromobenzene. His results are in fair agreement with the above formula, but unfortunately the method, in one case at least (carbon tetrachloride and benzene), gave inaccurate results, and it is therefore impossible to place complete reliance on the experimental data. The critical pressures of these substances have been determined and those of the components of the mixtures are in no case equal.

Determinations by Dynamical Method.—The vapour pressures of mixtures may be determined by the dynamical

method; the still is kept at a uniform temperature by
means of a suitable bath, and the pressures are observed
under which ebullition takes place. This method has been
adopted by Lehfeldt (p. 75) and by Zawidski (p. 76), and
the latter experimenter finds that mixtures of ethylene dibromide with propylene dibromide give results in conformity
with the formula $P = \dfrac{mP_A + (100 - m)P_B}{100}$. The critical
pressures of these substances are not, however, known.

Again, the vapour pressures of mixtures may be determined indirectly from their boiling points under a series of
pressures. The boiling points of each mixture are mapped
against the pressures or, better, the logarithms of the
pressures, and from the curves so obtained the pressure at
any required temperature can be read off.

In order to obtain the complete vapour pressure curve
for two liquids at a given temperature, the boiling point
determinations for a considerable number of mixtures would
have to be carried out through a wide range of pressure,
but a less elaborate investigation is sufficient to ascertain
whether or not the above formula is applicable.

Suppose that we determine the boiling points of mixtures
containing, say, 25, 50 and 75 molecules per cent. of the
less volatile component, A, at a few pressures above and
below 760 mm., in order to ascertain the boiling points
under normal pressure with accuracy. If, then, we know
the vapour pressures of each component at the three temperatures, we may calculate the theoretical vapour pressure
of each mixture from the formula.

Chlorobenzene and Bromobenzene.—The boiling points
of three mixtures of chlorobenzene and bromobenzene
were determined in this way (9), and the theoretical vapour
pressures at these temperatures were then calculated. The
results are given in Table 4.

TABLE 4.

Molecular percentage of C_6H_5Br.	Observed boiling point.	Vapour pressures at $t°$.			Actual pressure, P'.	$\Delta.$ $P' - P.$
		$P_A.$ $C_6H_5Br.$	$P_B.$ $C_6H_5Cl.$	$P.$ Mixture (calculated).		
25·01	136·75°	452·85	862·95	760·4	760·0	− 0·4
50·00	142·16	526·25	992·30	759·3	760·0	+ 0·7
73·64	148·16	618·40	1153·00	759·3	760·3	+ 0·7
			Mean	**759·7**	Mean	**+0·3**

The differences are within the limits of experimental error.

The critical pressures of chlorobenzene and bromobenzene are equal, or nearly so, and it has been found that, when the two liquids are mixed in equimolecular proportions, there is no perceptible alteration of temperature or of volume, and it may therefore be concluded that $a_{1.2} = \sqrt{a_1 . a_2}$. The conditions specified by Van der Waals are therefore fulfilled, and in this case, at any rate, the formula $P = \dfrac{mP_A + (100 - m)P_B}{100}$ gives the vapour pressures accurately.

TABLE 5.

Substance.	Critical pressure.	$\Delta.$
Ethyl acetate	38·00 atm.	} 4·83
Ethyl propionate	33·17 ,,	
Toluene	41·6* ,,	} 3·5
Ethyl benzene	38·1* ,,	
n-Hexane	29·62 ,,	} 4·98
n-Octane	24·64 ,,	
Toluene (10)	41·6 ,,	} 8·5
Benzene (10)	50·1 ,,	
Benzene (11)	47·88 ,,	} 2·91
Carbon tetrachloride	44·97 ,,	
Methyl alcohol	78·63 ,,	} 15·67
Ethyl alcohol	62·96 ,,	

* As the critical pressures of both toluene and ethyl benzene have been determined by Altschul (10), it seems best to give his value for benzene in the comparison with toluene.

Other Mixtures of Closely Related Compounds.—It does not follow, however, that it is only when the critical pressures are equal that the formula is applicable, and it will be seen from Table 6 (p. 40) that the deviations may be exceedingly small when the liquids are closely related, but their critical pressures (Table 5) are widely different (12).

Influence of Chemical Relationship.—For the first three pairs of substances and the last in Table 6, the changes of volume and of temperature are exceedingly small, but the expansion and fall of temperature are very noticeable with benzene and toluene, and it may be remarked that the chemical relationship is not quite so close in this case, the benzene molecule being wholly aromatic, the toluene partly aromatic and partly aliphatic.

With regard to the mean differences $P' - P$, it may be said that they are very small, the corresponding temperature differences being only 0·17°, 0·18°, 0·20°, 0·12° and 0·02° respectively, and the values of $100(P' - P)/P$ **0·58, 0·57, 0·60, 0·35,** and **0·08** [compare Table 7, p. 42]. The conclusion may be drawn from the results that for mixtures of closely related compounds the relation between vapour pressure and molecular composition is represented by a line which is very nearly, if not quite, straight.

Errors of Experiment.—The greater the difference between the boiling points of the two liquids, the more difficult is it to determine the boiling point of a mixture with accuracy, and the greater also to some extent is the probability of error in the value of P. This may partly account for the irregular results with n-hexane and n-octane.

Comparison of Heat Change, Volume Change and Vapour Pressure.—It does not appear that change of volume, or temperature, when the liquids are mixed can be relied on as a safe guide to the behaviour of a mixture as regards

TABLE 6.

Mixture.	Molecular percentage of A = m.	Pressures.			Change on mixing in equimolecular proportions.	
		Actual P'.	Calculated P.	Δ. P' − P.	Volume per cent.	Temperature.
		mm.	mm.	mm.		*
A. Ethyl propionate ⎫	25·01	760	756·5	+ 3·5		
⎬	50·00	,,	755·7	+ 4·3	+0·015	− 0·02°
B. Ethyl acetate . ⎭	74·62	,,	754·6	+ 5·4		
	Mean		755·6	+ 4·4		
A. Ethyl benzene . ⎫	25·02	760	762·8	− 2·8		
⎬	49·97	,,	763·5	− 3·5	− 0·034	+0·05
B. Toluene . . . ⎭	75·00	,,	765·5	− 5·5		
	Mean		763·9	− 3·9		
A. n-Octane . . . ⎫	23·31	760	752·0	+ 8·0		
⎬	50·00	,,	760·2	− 0·2	− 0·053	+0·06
B. n-Hexane . . . ⎭	74·99	,,	781·6	− 21·6		
	Mean		764·6	− 4·6		
A. Toluene . . . . ⎫	24·94	760	760·6	− 0·6		
⎬	50·00	,,	762·9	− 2·9	+0·161	− 0·45
B. Benzene . . . ⎭	72·46	,,	764·5	− 4·5		
	Mean		762·7	− 2·7		
A. Ethyl alcohol . ⎫						
⎬	50·13	760	759·4	+ 0·6	+0·004	−0·10
B. Methyl alcohol ⎭						

* The temperature changes given here are merely comparative; they were observed by mixing together, in a round-bottomed flask, equimolecular quantities (22 c.c. in all) of the two substances. The temperature given is the difference between that of the mixture and the mean of those of the components which never differed by more than 0·2°. The alterations of temperature show the direction and give a rough indication of the magnitude of the corresponding heat changes.

its vapour pressure, for one would expect that, as a general rule, the actual pressure should be lower than the calculated when there is contraction and rise of temperature, as, in fact, is observed with the second and third pairs of liquids; and that, when there is expansion and fall of temperature, the observed pressure should be higher than the calculated, as is the case with the first pair. With benzene and toluene, however, the rule is not followed, for the actual pressure is the lower, although there is expansion and fall of temperature on mixing.

Components not Closely Related.—When we come to consider the behaviour of liquids which are not so closely related, we find in many cases much greater changes of volume and temperature, and also much larger values for $100(P' - P)/P$, as will be seen from Table 7 (p. 42).

It will be seen that contraction is frequently accompanied by absorption of heat, and that $P' - P$ may have a positive value when there is contraction, and even when there is both contraction and evolution of heat on mixing the components. In the case of mixtures of the alcohols with benzene, it is remarkable that, although the value of $(P' - P)/P$ diminish as the molecular weights of the alcohols increase, yet the fall in temperature, from methyl to isobutyl alcohol, becomes greater, and the very small volume-changes pass from negative to positive.

General Conclusions.—The following conclusions may be drawn from the foregoing data :—

1. When the two components are chemically very closely related, the changes of volume and of temperature on mixing the liquids are, as a rule at any rate, inconsiderable.

2. When not only is the chemical relationship very close, but the critical pressures are equal and $a_{1.2} = \sqrt{a_1 \cdot a_2}$ (Van der Waals), the vapour pressure of the mixture is accurately represented by the formula $P = \dfrac{mP_A + (100 - m)P_B}{100}$.

3. When the components are very closely related, but the critical pressures are not equal, the percentage difference between the observed and calculated pressures—

TABLE 7.

Mixture.	Molecular percentage of $A = m$.	Pressures.			Changes on admixture.			
		Actual P'	Calculated P	$100 \frac{(P' - P)}{P}$	m	Volume per cent.	m	Temperature.
A. Ethylene chloride B. Benzene	50·0	252·4	252·1	+ 0·1	50·0	+0·34	50	- 0·35°
A. Benzene. B. Carbon tetrachloride	50·0	760·0	739·85	+ 2·7	50·0	-0·13	50	- 0·69
A. Toluene. B. Carbon tetrachloride	50·0	196·0	201·6	- 2·8	50·0	-0·07	50	+ 0·25
A. Ethyl acetate. B. Carbon tetrachloride	50·0	314·6	293·2	+ 7·3	50·0	+0·03	50	+ 0·55
A. Chlorobenzene B. Carbon tetrachloride	50·0	82·0	76·0	+ 7·9	50·0	-0·12	50	- 0·4
A. Benzene. B. n-Hexane	50·0	760·0	684·2	+11·1	50·0	+0·52	50	- 4·7
A. Carbon disulphide. B. Methylal	50·0	702·0	551·1	+27·4	49·7	+1·22	50	- 6·5
A. Acetone. B. Carbon disulphide	50·0	646·75	428·05	+51·1	49·9	+1·21	50	- 9·85
A. Chloroform. B. Acetone.	50·0	254·0	318·8	- 20·3	50·0	-0·23	50	+12·4
A. Water B. Methyl alcohol	50·0	282·0	247·0	+14·2	40·0	-2·98	40	+ 7·85
A. Water B. Ethyl alcohol	50·0	372·0	285·7	+30·2	40·0	-2·56	40	+ 2·95
A. Water B. Propyl alcohol	50·0	576·0	387·0	+48·8	40·0	-1·42	40	- 1·15
A. Isobutyl alcohol. B. Water	50·0	792·0	476·5	+66·2	60·0	-0·90	60	- 3·15
A. Benzene. B. Methyl alcohol	50·0	760·0	475·0	+60·0	50·0	-0·01	50	- 3·8
A. Benzene. B. Ethyl alcohol.	50·1	760·0	507·3	+49·8	55·0	0·00	50	- 4·2
A. Propyl alcohol. B. Benzene.	49·0	760·0	535·4	+41·95	50·0	+0·05	50	- 4·65
A. Isobutyl alcohol. B. Benzene.	50·0	760·0	562·2	+35·2	50·0	+0·16	50	- 6·35
A. Isoamyl alcohol. B. Benzene.	50·0	760·0	579·6	+31·1	50·2	+0·23	50	- 5·25

The pressures are from observations by Zawidski, Konowaloff, Lehfeldt, Kohnstamm, Jackson and Young, Fortey and Young, or have been specially determined.

$100(P' - P)/P$—is very small, even when, as with methyl and ethyl alcohol, there is molecular association in the liquid state.

4. When the two liquids are not closely related, even if there is no molecular association in the case of either of them, the percentage differences are, as a rule, much greater.

5. When the components are not very closely related and the molecules of one or both of them are associated in the liquid state, the values of $100(P' - P)/P$ are usually very large indeed.

Importance of Chemical Relationship and Molecular Association.—In considering, then, the probable behaviour, as regards vapour pressure, of two liquids, it would appear that chemical relationship is the chief point to be considered, and that, if the two liquids are not closely related, the question whether the molecules of either or both of them are associated in the liquid state is also of great importance.

Alcohols—Water—Paraffins.—The influence of chemical relationship is well seen by comparing the properties of mixtures of the monhydric aliphatic alcohols with water on the one hand and with the corresponding paraffins on the other (13).

These alcohols may be regarded as being formed from water by the replacement of one atom of hydrogen by an alkyl group, thus (C_nH_{2n+1})—O—H, or as hydroxyl derivatives of the paraffins, thus $C_nH_{2n+1}(OH)$.

The alcohols are thus related to, and their properties are intermediate between those of water and of the paraffins ; and it is found that, in the homologous series, as the magnitude of the alkyl group increases, the properties of the alcohols recede from those of water and approach those of the corresponding paraffins. This is well seen in Table 8 (page 44).

TABLE 8.

Number of Carbon Atoms.	Boiling points.				
	Paraffin.	Δ	Alcohol.	Δ	Water.
1	− 164 °	+ 228·7°	+ 64·7°	− 35·3°	100·0°
2	− 93	171·3	78·3	− 21·7	,,
3	− 45	142·4	97·4	− 2·6	,,
4	+ 1	116·0	117·0	+ 17·0	,,
5	36·3	101·7	138·0	38·0	,,
6	69·0	88·0	157·0	57·0	,,
7	98·4	77·6	176·0	76·0	,,
8	125·6	70·4	196·0	96·0	,,
16	287·5	56·5	344·0	244·0	,,

Thus, methyl alcohol boils only 35·3° lower than water, but 228·7° higher than methane, while cetyl alcohol boils 244° higher than water, but only 56·5° higher than the corresponding paraffin.

Water and Alcohols—Action of Dehydrating Agents. —Again, most dehydrating agents, which react or combine with water, behave in a somewhat similar manner towards the alcohols, though to a smaller degree, and to a diminishing extent as the molecular weight increases, and this fact accounts for the unsatisfactory results obtained with them. Thus, phosphoric anhydride gives phosphoric acid with water and a mixture of ethyl hydrogen phosphates with ethyl alcohol; with barium oxide, water forms barium hydrate, while ethyl alcohol forms, according to Forcrand, a compound $3BaO,4C_2H_6O$; sodium acts in precisely the same way on the alcohols as on water, but the intensity of the action diminishes rapidly as the complexity of the alkyl group increases; calcium chloride forms a crystalline hexahydrate with water and a crystalline tetra-alcoholate with methyl or ethyl alcohol; anhydrous copper sulphate dissolves rapidly in water, and, on evaporation, crystals of $CuSO_4,5H_2O$ are deposited; in methyl alcohol the sulphate dissolves slowly, but to a considerable extent, giving a blue solution from which, according to Forcrand, greenish-blue crystals of $CuSO_4, CH_4O$ may be obtained; anhydrous copper sulphate is, however, quite insoluble in ethyl alcohol, and will extract some water from strong spirit, but it is not a sufficiently powerful dehydrating agent to remove the whole.

Miscibility.—The gradual change in properties of the alcohols is also well shown by the fact that the lower alcohols are miscible with water in all proportions, the intermediate ones within limits, while the higher alcohols are practically insoluble in water.

Volume and Heat Changes.—Lastly, the contraction and heat evolution that take place on mixing the alcohols with water diminish with rise of molecular weight; there is, in fact, increasing heat absorption in the case of the higher alcohols.

Vapour Pressures.—The lowest member of the series, methyl alcohol, bears the closest resemblance to water, and, as Konowaloff has shown, the vapour pressures of mixtures of these substances are in all cases intermediate between those of the pure components, and the curve representing the relation between vapour pressure and molecular composition does not deviate very greatly from a straight line, as will be seen from Fig. 14, in which the vapour pressures of mixtures of four alcohols with water are given, at the boiling points of the alcohols under a pressure of 400 mm.

With ethyl alcohol and water the deviation is considerable, and more recent and accurate observations have shown that a particular mixture exerts a maximum vapour pressure, but the experiments of Konowaloff are not sufficiently numerous to bring out this point.

When, however, we come to mixtures of *n*-propyl alcohol and water, we find that there is a very well defined maximum pressure, although the liquids are miscible in all

proportions, and the curve shows considerable resemblance
to that representing the behaviour of the partially miscible
liquids, isobutyl alcohol and water.

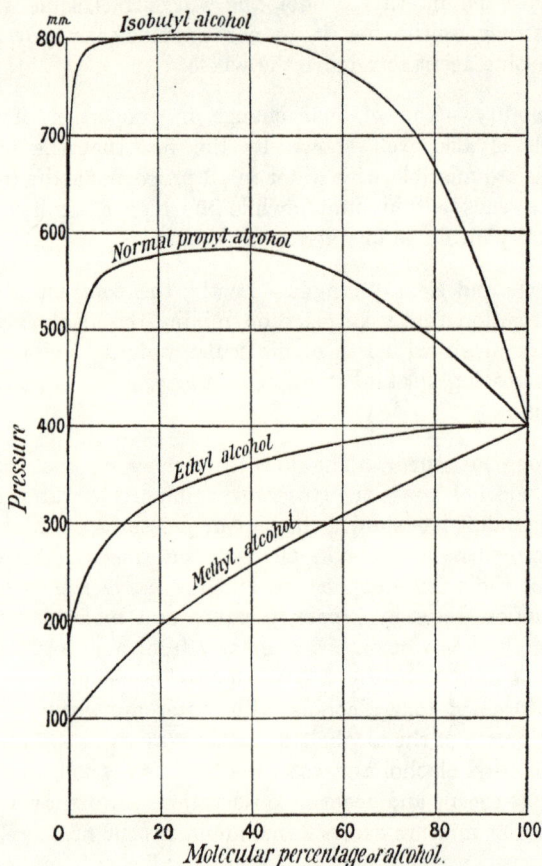

FIG. 14.—Vapour pressures of mixtures of alcohols with water.

Mixtures of Maximum Vapour Pressure.—The question whether a mixture of maximum vapour pressure will be formed in any given case depends partly on the deviation

of the vapour pressure-molecular composition curve from
straightness, partly on the difference between the vapour
pressures of the two components. Thus, in the case
of ethyl alcohol and water, the difference between the
vapour pressures at 63° is 229 mm., and there is a maximum
vapour pressure, very slightly higher than that of pure ethyl
alcohol, for a mixture containing about 89 or 90 molecules

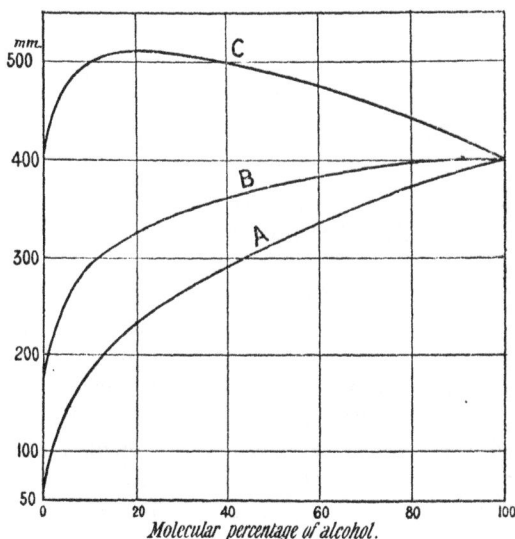

Fig. 15.

per cent. of alcohol (B, Fig. 15). If the difference between
the vapour pressures of the pure substances at the same
temperature were 350 mm., and the deviation of the curve
from straightness were the same, there would be no
maximum vapour pressure (A, Fig. 15); while, if the vapour
pressures of alcohol and water were equal and the deviation
the same, the maximum vapour pressure would be far more
obvious, and the molecular percentage of alcohol in the

mixture that exerted it would be something like 20, instead of about 90 (c, Fig. 15).

The gradual divergence of the properties of the alcohols from those of water, as the molecular weight increases, is indicated by the increasing curvature of the pressure-molecular composition lines (Fig. 14). The maximum differences between the pressures represented by the actual curves and the theoretical straight lines are roughly as follows :—

Methyl alcohol and water . . . 43 mm.
Ethyl . . . . 112 ,,
n-Propyl . . . . 203 ,,
Isobutyl . . . . 315 ,,

Mixtures of *n*-Hexane with Alcohols.—We cannot well study the miscibility of the alcohols with the corresponding paraffins, because the first four of these hydrocarbons are gaseous at the ordinary temperature, and most of them are very difficult to prepare in a pure state. Normal hexane, however, may be conveniently prepared by the action of sodium on propyl iodide, and is easily purified. This paraffin is only partially miscible with methyl alcohol at the ordinary temperature, but mixes in all proportions with those of the higher alcohols which have been investigated—ethyl, propyl, isobutyl and isoamyl alcohols. Again, normal hexane forms mixtures of maximum vapour pressure with the lower alcohols, but not with isoamyl alcohol or any other of higher boiling point. Lastly, the fall of temperature, on mixing hexane with the alcohols in equimolecular proportions, diminishes slightly as the molecular weights of the alcohol rise. [Ethyl alcohol − 2·55°, *n*-Propyl alcohol − 2·40°, Isobutyl alcohol − 2·35°, Isoamyl alcohol − 1·85°.] In all these respects, therefore, it may be said that the properties of the alcohols approach those of normal hexane, but recede from those of water, as their molecular weights increase.

Mixtures of Benzene with Alcohols.—Benzene is much more easily obtained in quantity than hexane, and behaves in a somewhat similar manner. The lower alcohols are miscible with benzene in all proportions; but while methyl, ethyl, isopropyl, n-propyl, tertiary butyl and isobutyl alcohols form mixtures of maximum vapour pressure with that hydrocarbon, isoamyl alcohol does not, and it is practically certain that no alcohol of higher boiling point would form such a mixture.

It will be seen from Table 7 that while, for mixtures of the alcohols with water, the values of 100 $(P' - P)/P$ show a steady rise as the molecular weights of the alcohols increase, the corresponding values for mixtures with benzene fall regularly. It may also be mentioned that the solubility of the alcohols in benzene, relatively to that in water, becomes greater as the molecular weights of the alcohols increase, for while methyl alcohol can be separated from its benzene solution with the greatest ease, and ethyl alcohol without difficulty by treatment with water, the extraction of propyl alcohol is more troublesome, and that of isobutyl or amyl alcohol by this process is exceedingly tedious. On the other hand, the variations in the temperature and volume changes, which occur on admixture are, generally, in the same direction for benzene as for water, though they are much smaller. It should, however, be noted that the relationship of the alcohols to the aromatic hydrocarbon, benzene, is by no means so close as to a paraffin, such as normal hexane; indeed, there is expansion and absorption of heat when benzene and hexane are mixed together and the two substances, in all probability, form a mixture of maximum vapour pressure.

REFERENCES

1. D. Berthelot, "On Mixtures of Gases," *Compt. rend.*, 1898, **126**, 1703.
2. Galitzine, "On Dalton's Law," *Wied. Ann.*, 1890, **41**, 770.

E

3. Guthrie, "On some Thermal and Volume Changes attending Admixture," *Phil. Mag.*, 1884, [V.], **18**, 495.
4. Speyers, "Some Boiling Point Curves," *Amer. Journ. Sci.*, 1900, IV., **9**, 341.
5. Van der Waals, "Properties of the Pressure Lines for Co-existing Phases of Mixtures," *Proc. Roy. Acad. Amsterdam*, 1900, **3**, 170.
6. Kohnstamm, "Experimental Investigations based on the Theory of Van der Waals," *Inaugural Dissertation, Amsterdam*, 1901.
7. F. D. Brown, "On the Volume of Mixed Liquids," *Trans. Chem. Soc.*, 1881, **39**, 202.
8. Linebarger, "On the Vapour Tensions of Mixtures of Volatile Liquids," *Journ. Amer. Chem. Soc.*, 1895, **17**, 615 and 690.
9. Young, "The Vapour Pressures and Boiling Points of Mixed Liquids, Part I.," *Trans. Chem. Soc.*, 1902, **81**, 768.
10. Altschul, "On the Critical Constants of some Organic Compounds," *Zeitschr. physik. Chem.*, 1893, **11**, 577.
11. Young, "On the Vapour Pressures and Specific Volumes of similar Compounds of Elements in Relation to the Position of those Elements in the Periodic Table," *Trans. Chem. Soc.*, 1889, **55**, 486.
12. Young and Fortey, "The Vapour Pressures and Boiling Points of Mixed Liquids, Part II.," *Trans. Chem. Soc.*, 1902, **83**, 45.
13. Young and Fortey, "The Properties of Mixtures of the Lower Alcohols (1) with Water, (2) with Benzene, and with Benzene and Water," *Trans. Chem. Soc.*, 1902, **81**, 717 and 739.

CHAPTER IV

Statical and Dynamical Methods of Determination.
—It has been stated that the boiling point of a pure liquid
under a given pressure may be determined by either the
statical or the dynamical method, the curve which shows
the relation between temperature and pressure representing
not only the vapour pressures of the liquid at different temperatures, but also its boiling points under different pressures ; this statement applies equally to the boiling points
of any given mixture under different pressures.

As regards the vapour pressures of mixtures, it has been
shown that there are two simple cases : —

a. That of non-miscible liquids, for which the vapour
pressure of the two (or more) liquids together is equal to
the sum of the vapour pressures of the components at the
same temperature ;

b. That of closely related compounds, which are miscible
in all proportions, and of a few other pairs of infinitely
miscible liquids, for which the formula

$$P = \frac{mP_A + (100 - m)P_B}{100}$$

holds good. It is in these two cases only that the boiling
points can be calculated from the vapour pressures of the
components.

Non-Miscible Liquids

Dalton's law of partial pressures is applicable to the case
of non-miscible liquids, each vapour behaving as an indiffer-

ent gas to the others, and the boiling point of each liquid depends on the partial pressure of its own vapour. The temperature is necessarily the same for all the liquids present, and the total pressure, if the distillation is carried out in the ordinary way, is equal to that of the atmosphere. The boiling point is therefore that temperature at which the sum of the vapour pressures of the components is equal to the atmospheric pressure.

Calculation from Vapour Pressures (Chlorobenzene and Water).

—In Table 9 the vapour pressures of chlorobenzene and of water (1), two liquids which are practically non-miscible, are given for each degree from 89° to 93°, and also the sum of the vapour pressures at the same temperatures.

TABLE 9.

Vapour Pressures in mm.

Temperature.	Chlorobenzene.	Water.	Sum.
89°	201·15	505·75	706·9
90	208·35	525·45	733·8
91	215·8	545·8	761·6
92	223·45	566·75	790·2
93	231·3	588·4	819·7

Thus, when chlorobenzene and water are heated together, say in a barometer tube, the observed vapour pressure at 90° should be 733·8 mm., 761·6 mm. at 91°, and so on. Conversely, when chlorobenzene and water are distilled together in an ordinary distillation bulb, so that the vapours are unmixed with air, the observed boiling point, when the atmospheric pressure is 761·6 mm., should be 91°.

Experimental Verification.

—In an actual experiment, 100 c.c. of chlorobenzene and 80 c.c. of water were distilled together when the barometric pressure was 740·2 mm., and it was found that the temperature varied only between 90·25° and 90·35°, until there was scarcely any chlorobenzene visible in the residual liquid, when it rose rapidly to nearly

100°. The theoretical boiling point is calculated as follows :—
The increase of pressure for 1° rise of temperature from 90° to
91° is $761 \cdot 6 - 733 \cdot 8 = 27 \cdot 8$ mm. It may be assumed without
sensible error that the value of dp/dt is constant over this small
range of temperature and the boiling point should therefore
be $\left(90 + \dfrac{740 \cdot 2 - 733 \cdot 8}{761 \cdot 6 - 733 \cdot 8} \right)^{\circ} = 90 \cdot 23°$, which is very close to
that actually observed.

LIQUIDS MISCIBLE WITHIN LIMITS

Before considering the behaviour of those infinitely
miscible liquids for which the relation between vapour
pressure and molecular composition is represented by a
straight line, it will be well to take the case of partially
miscible liquids.

The boiling point of a pair of such substances is higher
than that calculated as above for non-miscible liquids, but if
the miscibility is slight, the difference between the observed
and calculated temperatures is not serious, and the observed
boiling point will be decidedly lower than that of either
component.

Aniline and Water.—Water dissolves only about 3 per
cent. of aniline, and aniline about 5 per cent. of water
at 12°, though the solubility in each case is considerably
greater at 100°. Fifty c.c. of aniline and 200 c.c. of water
were distilled together under a pressure of 746·4 mm., and
the boiling point was found to remain nearly constant at
98·75° for some time but afterwards rose slowly to 99·65°
while there was still a moderate amount of water visible
with the residual aniline. The distillation was then stopped
and a large amount of water was added to the residue ; the
temperature, when the distillation was recommenced, rose
to 98·9°. It would therefore appear that the composition of
the liquid in the still may alter considerably without pro-
ducing any great change in the boiling point, though the

temperature does not remain so constant as it does with chlorobenzene and water.

The vapour pressures of aniline and water at 97° to 99° are as follows :—

TABLE 10.

Vapour Pressures in mm.

Temperature.	Aniline.	Water.	Sum.
97°	40·5	682·0	722·5
98	42·2	707·3	749·5
99	44·0	733·3	777·3

The calculated boiling point under a pressure of 746·4 mm. would therefore be $\left(97 + \dfrac{746\cdot4 - 722\cdot5}{749\cdot5 - 722\cdot5} \right)°$ = 97·9°, which is 0·85° lower than the temperature which remained nearly constant for some time and 1·75° lower than that which was observed when there was still a moderate amount of water present.

Practical Application.—Advantage is taken of the fact that the boiling point of a pair of non-miscible or slightly miscible liquids is lower than that of either pure component, to distil substances which could not be heated to their own boiling points without decomposition, or which are mixed with solid impurities.

As a rule, water is the liquid with which the substance is distilled and the process is commonly spoken of as "steam distillation."

As an example, the commercial preparation of aniline may be described. Nitrobenzene is reduced by finely divided iron and water with a little hydrochloric acid, the products formed being ferrous chloride, magnetic oxide of iron and aniline. The aniline is distilled over with steam and the greater part of it separates from the distillate on standing. The aqueous layer, which contains a little aniline, is afterwards placed in the boiler which supplies steam to the still, so that in the next distillation the aniline is carried over into the still again.

The apparatus is shown diagrammatically in Fig.
16. A is the pipe by which steam is introduced into
the still, and B is a rotating
stirrer.

LIQUIDS MISCIBLE IN ALL PROPORTIONS

The boiling point of a mixture
of two liquids which are miscible
in all proportions can be calculated if the vapour pressure of
the mixture agrees accurately
with that given by the formula
$P = \dfrac{mP_A + (100 - m)P_B}{100}$. The
boiling points of such mixtures

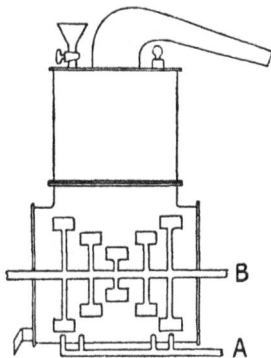

Fig. 16.—Aniline still.

are, however, not so simply related as their vapour pressures to the molecular composition.

Calculation from Vapour Pressure and Composition.

—In order to calculate the boiling points of all mixtures of two such liquids under normal pressure, we should
require to know the vapour pressures of each substance at
temperatures between their boiling points. Thus chlorobenzene boils at 132·0° and bromobenzene at 156·1°, and we
must be able to ascertain the vapour pressures of each
substance between 132° and 156°.

The percentage molecular composition of mixtures which
would exert a vapour pressure of 760 mm. must then be calculated at a series of temperatures between these limits, say
every two degrees, by means of the formula $m = 100\dfrac{P_B - P}{P_B - P_A}$,
where, in this case, $P = 760$.

Lastly, the molecular percentages of A, so calculated, must
be mapped against the temperatures, and the curve drawn
through the points will give us the required relation between

boiling point and molecular composition under normal pressure.

Closely Related Liquids.—For such closely related substances as chlorobenzene and bromobenzene, the ratio of the boiling points on the absolute scale is a constant at all equal pressures, as is also the value of $T \cdot dp/dt$, and it appears, so far as experimental evidence goes, that the ratio of the boiling point (abs. temp.) of any given mixture to that of one of the pure substances at the same pressure is also a constant at all pressures ; it would therefore, strictly speaking, be sufficient to determine the vapour pressures of either substance through a wide range of temperature and the ratios of the boiling points (abs. temp.) of the other pure substance and of a series of mixtures to that of the standard substance at a single pressure, in order to be able to calculate the boiling point of any mixture at any pressure.

The vapour pressures of chlorobenzene and bromobenzene from 132° to 156° are given in the table below, also the values of $P_B - 760$, $P_B - P_A$ and m.

TABLE 11.

Temperature.	Vapour Pressures in mm.				
	P_B Chloro-benzene.	P_A Bromo-benzene.	$P_B - 760$	$P_B - P_A$	m
132°	760·25	395·1	0·25	365·15	0·07
134	802·15	418·6	42·15	383·55	10·99
136	845·85	443·2	85·85	402·65	21·32
138	891·4	468·9	131·4	422·5	31·10
140	938·85	495·8	178·85	443·05	40·37
142	988·2	523·9	228·2	464·3	49·15
144	1039·5	553·2	279·5	486·3	57·47
146	1092·9	583·85	332·9	509·05	65·40
148	1148·4	615·75	388·4	532·65	72·92
150	1206·0	649·05	446·0	556·95	80·08
152	1265·8	683·8	505·8	582·0	86·91
154	1327·9	719·95	567·9	607·95	93·41
156	1392·3	757·55	632·3	634·75	99·61

In the diagram (Fig. 17) the values of m have been plotted against the temperatures, and the curve is drawn through the points which are not themselves indicated.

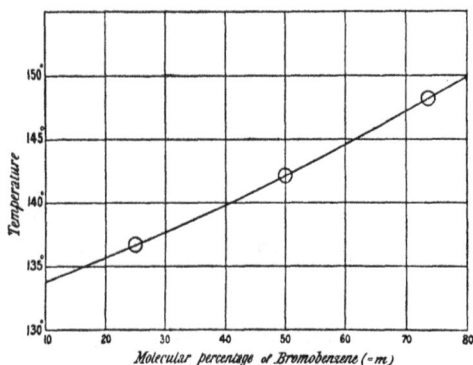

Fig. 17.—Boiling points of mixtures of bromobenzene and chlorobenzene.

Experimental Verification.—The boiling points of three mixtures of chlorobenzene and bromobenzene were determined at a series of pressures between about 690 and 800 mm. (2). The logarithms of the pressures were plotted against the temperatures, and the boiling points under normal pressure were read from the curve with the following results :—

Molecular percentage of bromobenzene . 25·01   50·00   73·64
Boiling point . . . . 136·75° 142·16° 148°·16

The points representing these values are indicated in Fig. 17 by circles, and it will be seen that they fall very well indeed on the theoretical curve.

Determination by Dynamical Method.—In determining the boiling points of mixed liquids by the dynamical method, it is of great importance that the vapour phase should be as small as possible, because if a relatively large amount of liquid were converted into vapour the composition of the residual liquid would, as a rule, differ sensibly from that of the original mixture. A reflux condenser must be

used and it is advisable that the thermometer should be
shielded from any possible cooling effect of the returning
liquid.   The temperature of the vapour should be read,
that of a boiling liquid, even when pure, being higher than
the "boiling point" (p. 24); but as the difference between
the temperatures of liquid and vapour is practically constant
for a given mixture, in a given apparatus with a steady
supply of heat, it is a good plan to read both temperatures
at each pressure, to subtract the average difference between
them from the temperature of the liquid, and to take the
mean of the read temperature of the vapour and the reduced
temperature of the liquid at each pressure as the true boiling
point.

Apparatus.—A suitable apparatus (2) is shown in Fig.
18.   It consists of a bulb of about 155 c.c. capacity with

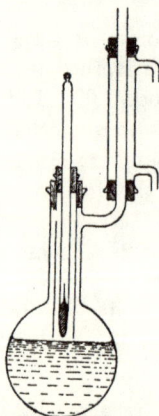

FIG. 18.—Boiling point
apparatus for mixtures.

a wide vertical tube, to which is sealed
a narrow side tube cooled by water to
act as a reflux condenser.   The upper
end of the side tube is connected with
an exhaust and compression pump and
a differential gauge.   The wide vertical
tube is provided with a well-fitting cork,
through which passes a rather narrower
thin-walled tube, which has a hole blown
in it just below the cork.   This narrower
tube is also fitted with a cork, through
which passes the thermometer.

The quantity of each mixture placed
in the bulb was such that its volume at
the boiling point was about 125 c.c.; the
volume of vapour in the bulb and verti-
cal tubes was about 75 c.c.   The thin-walled tube was
pushed down until the bottom of it was about 3 mm. above
the surface of the liquid when cold, and the bottom of the
thermometer bulb was about level with that of the tube.

This arrangement possesses the following advantages :

a. The liquid that returns from the reflux condenser cannot come near the thermometer, and the amount of liquid that condenses on the thermometer, and on the inner walls of the thin-walled tube, is exceedingly small; on the other hand, with the large quantity of liquid present and the small flame that is required, there is no fear of the vapour being super-heated.

b. It is possible to take readings of the temperature both of the vapour and of the liquid without altering the position of the thermometer, for when the burner is directly below the centre of the bulb, the liquid boils up into the thin-walled tube well above the thermometer bulb, but when the burner is moved a little to one side, the surface of the liquid immediately below that tube remains undisturbed and the liquid does not come in contact with the thermometer bulb.

Calculated Boiling Point-Molecular Composition Curves.—The relation between the boiling points and the molecular composition of mixtures of chlorobenzene and bromobenzene is represented by a curve (Fig. 17), the temperatures being lower than if the line were straight, and that is the case for any other pair of liquids for which the formula $P = \dfrac{mP_A + (100 - m)P_B}{100}$ holds good. It is found that the curve approximates more and more closely to a straight line as the difference, Δ, between the boiling points of the components diminishes. If the relation between boiling point and molecular composition were represented by a straight line, the boiling point of any mixture could be calculated from the formula $t = \dfrac{mt_A + (100 - m)t_B}{100}$ or, more conveniently, $t = t_B + \dfrac{m}{100}(t_A - t_B)$, and it is useful to consider the deviations of the theoretical boiling point curves from straightness (3). The boiling point of propyl acetate is 101·58° and of methyl acetate 57·15°, and $t_A - t_B = \Delta = 44\cdot43°$. The vapour pressures of both substances at temperatures between 57° and 102° are known, and in Table 12 are given the molecular percentages, m, of propyl acetate calculated from the formula $m = \dfrac{100(P_B - 760)}{P_B - P_A}$ for each fourth degree from 58° to 98°, the straight line temperatures for these molecular percentages calculated from the formula $t' = t_B + \dfrac{m\,\Delta}{100}$, and the differences $d = t - t'$.

TABLE 12.

t	m	t'	d
58°	3·67	58·78°	− 0·78
62	19·38	65·76	− 3·76
66	32·94	71·78	− 5·78
70	44·69	77·01	− 7·01
74	54·95	81·57	− 7·57
78	63·95	85·56	− 7·56
82	71·89	89·09	− 7·09
86	78·94	92·22	− 6·22
90	85·21	95·01	− 5·01
94	90·83	97·51	− 3·51
98	95·88	99·75	− 1·75

Maximum Deviation from Straight Line.—Plotting d against m, we get the curve shown in Fig. 19, from the form of which it is evident that the maximum deviation, D_1, can be ascertained with accuracy and the corresponding molecular percentage, m_D, approximately. With other pairs of liquids, the difficulty in estimating the true value of m_{D1} increases as D_1 diminishes. On determining D_1 and m_{D1} for nineteen pairs of closely related liquids, paraffins, aromatic hydrocarbons, halogen derivatives of benzene, esters, and alcohols, it was found that in all cases the values of m_{D1} were given very satisfactorily by the formula

$$m_{D1} = 50 + 0\cdot18\,\Delta.$$

For any two of the four mono-halogen derivatives of benzene, the values of D_1 agree well with those given by the formula $D_2 = \dfrac{K \Delta^2}{T'}$, where K is a constant and

$$T' = \frac{(50 + 0.18\,\Delta)T_A + (50 - 0.18\,\Delta)T_B}{100},$$ but for other

pairs of closely related liquids which have different values of $c \left(= \dfrac{dt}{dp} \cdot \dfrac{1}{T} \right)$, the formula $D_2 = \dfrac{K'.\Delta^2.c_A}{c'.T'.c_B}$, must be used, where $K' = -0.000158$ and

$$c' = \frac{(50 + 0.18\,\Delta)c_A + (50 - 0.18\,\Delta)c_B}{100}$$

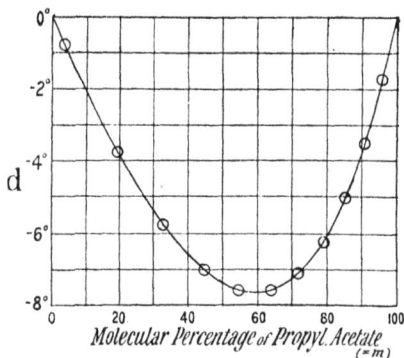

FIG. 19.—Deviation of boiling point curve from straight line.

For the nineteen pairs of liquids the greatest difference between D_2, calculated from the above formula, and D_1, calculated from the vapour pressures, is only $0.16°$, although the actual values of D_1 vary from $-1.62°$ to $-32.73°$ (3).

Components not Closely Related.—Nineteen pairs of liquids the members of which are not closely related were also investigated, and it was found that the differences between the values of m_{D1} calculated from the formula $m_{D1} = 50 + 0.18\,\Delta$ and those deduced from the vapour pressures are generally small when c_A/c_B does not differ greatly from unity; they are invariably positive when c_A/c_B is very small and negative when large.

As regards the maximum deviation, D_1, it must be remembered that when Δ has a low value, both D_1 and D_2 and therefore $D_2 - D_1$ are necessarily very small; the relative as well as the actual differences should therefore be considered.

It will be seen from Table 13 that when c_A/c_B is less than unity and Δ is small, D_1 may have a small positive value amounting, in the case of ethyl alcohol and n-hexane, to $0.23°$.

TABLE 13.

Liquids in mixture.		Boiling points absolute temp.		Δ	c_A/c_B	D_1	D_2	D_2-D_1
A.	B.	A.	B.					
Propyl alcohol . . .	n-Pentane	370·47	309·13	61·34	0·752	− 11·66	− 9·66	+2·00
Ethyl alcohol . . .	Carbon tetrachloride	351·32	349·83	1·49	0·764	+ 0·09	− 0·01	−0·10
Ethyl alcohol . . .	n-Hexane	351·32	341·97	9·35	0·770	+ 0·23	− 0·29	−0·52
Propyl alcohol . . .	Benzene	370·47	353·27	17·20	0·777	− 0·15	− 0·94	−0·79
*Water	Benzene	373·00	353·27	19·73	0·818	− 0·58	− 1·27	−0·69
Benzene	Carbon disulphide .	353·27	319·25	34·02	0·953	− 4·02	− 4·16	−0·14
Carbon tetrachloride	Carbon disulphide .	349·83	319·25	30·58	0·969	− 3·34	− 3·41	−0·07
Benzene	Carbon tetrachloride	353·27	349·83	3·44	0·984	− 0·02	− 0·04	−0·02
Benzene	n-Hexane	353·27	341·97	11·30	0·992	− 0·44	− 0·47	−0·03
Benzene	Ether	353·27	307·60	45·67	1·000	− 8·20	− 8·06	+0·14
Water	Methyl alcohol. . .	373·00	337·89	35·11	1·000	− 5·57	− 5·50	+0·07
Water	Ethyl alcohol . .	373·00	351·32	21·68	1·053	− 2·36	− 2·21	+0·15
Water	Propyl alcohol . .	373·00	370·47	2·53	1·053	− 0·06	− 0·08	+0·03
Benzene	Methyl alcohol. .	353·27	337·70	15·57	1·210	− 1·77	− 1·21	+0·56
*Toluene	Water	383·60	373·00	10·60	1·212	− 0·95	− 0·52	+0·43
Carbon tetrachloride	Methyl alcohol. . .	349·83	337·89	11·94	1·242	− 1·26	− 0·73	+0·53
n-Octane	Ethyl alcohol . . .	398·60	351·32	47·27	1·266	− 11·34	− 10·85	+0·49
Toluene	Ethyl alcohol . . .	383·60	351·32	32·28	1·277	− 6·09	− 5·24	+0·85
Benzene	Ethyl alcohol . . .	353·27	351·32	1·95	1·287	− 0·13	− 0·02	+0·11

* Non-miscible.

For values of c_A/c_B between 0·95 and 1·05 the differences, $D_2 - D_1$, are quite small and in only one case in the whole table does the difference exceed $1°$.

We may therefore conclude that the maximum temperature deviation which would be observed if the formula $P = \dfrac{mP_A + (100 - m)P_B}{100}$ held good, may be calculated with considerable accuracy from the formula

$$D_2 = -\frac{0 \cdot 000158 . \Delta^2 . c_A}{c' . T' . c_B}$$

when the liquids are closely related or, in other cases, when c_A/c_B has a value between 0·95 and 1·05 and probably between 0·9 and 1·1, and in the majority of cases within 1° whatever the value of c_A/c_B.

Comparison of Observed with Calculated Boiling Points. Closely Related Liquids.

—We must now consider whether the actual boiling points of mixed liquids agree with those calculated from the vapour pressures, and, in particular, whether the observed maximum deviations (D_3), agree with those deduced from the vapour pressures (D_1), or, when these are not available, from those calculated from the formula $D_2 = -\dfrac{0 \cdot 000158 . \Delta^2 . c_A}{c' . T' . c_B}$.

We have already seen that the agreement is very good in the case of chlorobenzene and bromobenzene, and data are also available for four other pairs of closely related liquids (4), and are given in Table 14.

TABLE 14.

Liquids in Mixture.		D_1	D_3	$D_3 - D_1$
A.	B.			
Bromobenzene . . .	Chlorobenzene . .	$- 1 \cdot 91°$	$- 1 \cdot 94°$	$- 0 \cdot 03°$
n-Octane	n-Hexane	$- 10 \cdot 80$	$- 10 \cdot 53$	$+ 0 \cdot 27$
Toluene	Benzene	$- 3 \cdot 22$	$- 3 \cdot 06$	$+ 0 \cdot 16$
Ethyl benzene . . .	Toluene	$- 2 \cdot 22$	$- 1 \cdot 99$	$+ 0 \cdot 23$
Ethyl propionate .	Ethyl acetate . .	$- 1 \cdot 83$	$- 2 \cdot 00$	$- 0 \cdot 17$

For mixtures of methyl and ethyl alcohol, the difference $D_3 - D_1$ is also exceedingly small.

With chlorobenzene and bromobenzene, which show no change of temperature or of volume when mixed together, and have the same critical pressure, and with methyl and ethyl alcohol, which have widely different critical pressures, the differences $D_3 - D_1$, are certainly within the limits of experimental error; in the other cases the differences are very small, and it may probably be concluded that the actual boiling points of mixtures of closely related substances may be calculated with very considerable accuracy from the vapour pressures by means of the formula $m = \dfrac{100(P_R - 760)}{P_B - P_A}$, and that the maximum temperature deviation, D, may be calculated with fair accuracy from the formula

$$D = -\frac{0 \cdot 000158 . \Delta^2 . c_A}{c' . T' . c_B}.$$

Liquids not Closely Related.—As regards liquids which are not closely related it is only in a comparatively small number of cases that D_3 can be accurately ascertained, experimental data being either insufficient or altogether wanting, but there is evidence that the actual boiling points of such mixtures are, as a rule, considerably lower than those calculated from the vapour pressures, though occasionally higher, and that the formation of a mixture of constant boiling point is by no means an uncommon occurrence.

Data, more or less complete, for eighteen liquids are given in the following table (3). Where the value of D_3 can only be approximately ascertained, the number is enclosed in a bracket. In a few cases, the boiling point and molecular composition of the mixture of constant boiling point only can be ascertained, and these values are given under d and m.

TABLE 15.

Liquids in mixture.		D_1	D_3	$D_3 - D_1$	Mixture of constant boiling point.	
A.	B.				d	m
Ethyl alcohol . . .	Carbon tetrachloride	+0·09°	...	...	Mixture formed.	
Ethyl alcohol . . .	n-Hexane	+0·23	...	...	-13·40	33·2
Propyl alcohol . . .	Benzene	-0·15	...	...	- 6·75	20·9
*Water	Benzene	-0·58	...	...	-16·85	29·6
Benzene	Carbon disulphide .	-4·02	- 8·0°	- 4·0°	} None formed.	
Carbon tetrachloride	Carbon disulphide .	-3·34	- 5·8	- 2·5		
Benzene	Carbon tetrachloride	-0·02	- 0·9	- 0·9	} Mixture probably	
Benzene	n-Hexane	-0·44	- 3·9	- 3·45	formed.	
Benzene	Ether	-8·20	...	...	} None formed.	
Water	Methyl alcohol. . .	-5·57	-10·5	-4·95		
Chloroform	Acetone	-0·04†	...	...	+ 4·9	80
Water	Ethyl alcohol . . .	-2·36	-12·26	- 9·9	- 2·46	10·6
Water	Propyl alcohol . . .	-0·06	(-11·4)	(-11·3)	-11·2	56·8
Benzene	Methyl alcohol	-1·77	...	...	-12·35	38·6
*Toluene	Water	-0·95	...	...	-20·3	44·4
Carbon tetrachloride	Methyl alcohol. . .	-1·26	...	...	-14·5	44·6
Toluene	Ethyl alcohol . . .	-6·09	...	...	Mixture formed.	
Benzene	Ethyl alcohol . . .	-0·13	...	...	-11·16	55·2

* Non-miscible.   † D_2.

Mixtures of Minimum Boiling Point.—In the great
majority of cases where the formation of mixtures of
minimum boiling point has been observed, one of the two
liquids is a hydroxyl compound—an alcohol, an acid or
water—and water also forms such mixtures with all the
lower alcohols, except methyl alcohol. It is well known
that the molecules of these liquids are more or less associated in the liquid state, and we may therefore conclude
that mixtures of minimum boiling point (maximum vapour
pressure) are most readily formed, and that $D_3 - D_1$ has the
greatest negative value when one or both of the liquids
exhibit molecular association.

It is probable that the molecules of acetone, and also of
the lower aliphatic esters, are associated to a slight extent
in the liquid state, and, according to Ryland (5), the following pairs of liquids form mixtures of minimum boiling
point: carbon disulphide and acetone; carbon disulphide
and methyl acetate; carbon disulphide and ethyl acetate;

F

acetone and methyl acetate; acetone and ethyl iodide; ethyl iodide and ethyl acetate.

Zawidski (6) has also shown that the first and last of these pairs of liquids form mixtures of maximum vapour pressure. According to Speyers (7) a mixture of minimum boiling point cannot be formed when both constituents have normal molecular weight at all concentrations, but this conclusion does not appear to be borne out by the facts, and it is clear that, when Δ is very small, a comparatively slight difference between D_3 and D_1 would be sufficient to account for the formation of a mixture of minimum boiling point. It is probable that benzene and n-hexane ($\Delta = 11\cdot3°$) and benzene and carbon tetrachloride ($\Delta = 3\cdot44°$) form such mixtures, and Ryland (*loc. cit.*) states that a mixture of carbon disulphide and ethyl bromide ($\Delta =$ about $7\cdot6°$), containing 32 per cent. by weight of the sulphide, boils $0\cdot5°$ lower than ethyl bromide. Again, Zawidski (*loc. cit.*) finds that methylal and carbon disulphide ($\Delta = 4\cdot15°$) form a mixture of maximum vapour pressure.

Mixtures of Maximum Boiling Point.—The occurrence of mixtures of maximum boiling point (or minimum vapour pressure) is comparatively rare; in most of the known cases one of the substances is an acid and the other a base or a compound of basic character—formic, acetic and propionic acid with pyridine (Zawidski) (6), hydrochloric acid with methyl ether (Friedel) (8); or the liquids are water and an acid—formic, hydrochloric, hydrobromic, hydriodic, hydrofluoric, nitric or perchloric acid (Roscoe) (9); but Ryland finds that such mixtures are formed by chloroform and acetone ($\Delta = 4\cdot8°$) and by chloroform and methyl acetate ($\Delta =$ about $4\cdot3°$); the first observation has been confirmed by Zawidski and by Kuenen and Robson (10), and the second by Miss Fortey (unpublished).

List of Known Mixtures of Constant Boiling Point.
—The occurrence of mixtures of maximum or minimum

boiling point is apt to give rise to mistakes in interpreting the results of a fractional distillation, and it may therefore be useful to give a list of the cases which have so far been observed.

TABLE 16.

Mixtures of Minimum Boiling Point—Binary mixtures.

| Substances in mixture. | | Boiling points. | | | Percentage of A by weight in mixture. | Observer. |
A.	B.	A.	B.	Mixture.		
Water	Ethyl alcohol	100·0°	78·3°	78·15°	4·43	{ N. & W., Y. & F.
Water	Isopropyl alcohol	100·0	82·45	80·35	12·10	Y. & F.
Water	Tert. butyl alcohol	100·0	82·55	79·9	11·76	,,
Water	n-Propyl alcohol	100·0	97·2	87·7	28·31	Y. & F., K.
Water	Allyl alcohol	100·0	...	...	...	K.
Butyric acid	Water	159·5	100·0	99·2	20	Ry.
Pyridine	Water	115	100·0	92·5	59	G. & C.
Benzene	Methyl alcohol	80·2	64·7	58·35	60·45	Y. & F.
Benzene	Ethyl alcohol	80·2	78·3	68·25	67·64	,,
Isopropyl alcohol	Benzene	82·45	80·2	71·9	33·3	,,
Tert. butyl alcohol	Benzene	82·55	80·2	73·95	86·6	,,
n-Propyl alcohol	Benzene	97·2	80·2	77·1	16·9	,,
Isobutyl alcohol	Benzene	108·05	80·2	79·85	9·3	,,
Allyl alcohol	Benzene	95·5	79·2	76·5	20	Ry.
Toluene	Ethyl alcohol	110·6	78·3	76·7	...	Y.
Toluene	n-Propyl alcohol	109	95·7	91·5	47	Ry.
Toluene	Isobutyl alcohol	109	105·8	100	57	,,
Toluene	Allyl alcohol	109	95·5	91·5	50	,,
Metaxylene	Amyl alcohol	136·5	128·5	125·5	48	,,
Paraxylene	Amyl alcohol	137·2	128·5	125·5	48	,,
Orthoxylene	Amyl alcohol	140·5	128·5	127·5	...	,,
n-Hexane	Methyl alcohol	68·95	64·7	50·0	...	Y.
Ethyl alcohol	n-Hexane	78·3	68·95	58·65	21·0	Y.
Propyl alcohol	n-Hexane	97·2	68·95	65·65	...	,,
Isobutyl alcohol	n-Hexane	108·05	68·95	68·1	...	,,
Acetic acid	Benzene	118·5	80·2	80·05	2	Ne.
Acetic acid	Toluene	117·5	109	104	30	Ry.
Metaxylene	Acetic acid	136·5	117·5	114	73	,,
Carbon tetrachloride	Methyl alcohol	76·75	64·7	55·7	79·4	Th., Y.
Methyl alcohol	Ethyl bromide	64·7	38·0	35·5	5	Ry.
Ethyl iodide	Methyl alcohol	72·9	64·7	55·0	83	,,
Methyl alcohol	Chloroform	64·7 / 64·8	60·5 / 61·45	54·0 / 53·5*	12 / 10	,, / P.
Methyl cyanide	Methyl alcohol	81·6	64·8	63·7	23	V. & D.
Isobutyl iodide	Methyl alcohol	118·5	64·7	64·0	...	Ry.
Ethyl alcohol	Ethyl bromide	77·8	38·0	37·0	...	,,
Ethyl alcohol	Ethyl iodide	77·8	72·0	63·0	14	,,
Ethyl alcohol	Chloroform	77·8	60·5	59·0	6	,,
Methyl cyanide	Ethyl alcohol	81·6	78·4	72·6	45	V. & D.
Isobutyl iodide	Ethyl alcohol	118·5	77·8	77·0	30	Ry.
Isopropyl alcohol	Ethyl iodide	81·5	72·0	66·0	66	,,
Isobutyl iodide	Isopropyl alcohol	118·5	81·5	81·5(?)	30	,,
n-Propyl alcohol	Ethyl iodide	95·7	72·4	70·0	7	,,
Isobutyl iodide	n-Propyl alcohol	118·5	95·7	93·0	55	,,
Amyl bromide	n-Propyl alcohol	118·2	95·5	94·0	29·3	H.
Amyl iodide	n-Propyl alcohol	146·5	95·7	95·6(?)	7 (?)	,,
Isobutyl iodide	Isobutyl alcohol	118·5	105·8	101·5	...	Ry.

TABLE 16 (*continued*).

Mixtures of Minimum Boiling Point—Binary mixtures (continued).

Substances in mixture.		Boiling points.			Percentage of A by weight in mixture.	Observer.
A.	B.	A.	B.	Mixture.		
Ethylene dibromide.	Isobutyl alcohol . .	129·5°	105·8°	105·0° :	38	Ry.
Amyl bromide . .	Isobutyl alcohol . .	118·1	105·0	103·4	36·4	H.
Amyl iodide . . .	Isobutyl alcohol . .	146·5	104·8	104·7(?)	5 (?)	,,
Amyl alcohol . . .	Isobutyl iodide . .	128·5	118·5	115·5	...	Ry.
Ethylene dibromide.	Amyl alcohol . . .	129·5	128·5	121·5	70	,,
Amyl alcohol . . .	Amyl bromide . .	129·0	117·9	116·15	12·7	H.
Amyl iodide . . .	Amyl alcohol . . .	146·5	128·9	127·3	48	,,
Butyric acid . . .	Bromobenzene. . .	159·5	152·5	147·5	19	Ry.
Methyl alcohol . .	Methyl acetate . .	64·7	56·0	54·0	18	,,
Ethyl alcohol . . .	Ethyl acetate . . .	77·8	76·0	71·5	31	,,
Isopropyl alcohol . .	Ethyl acetate . . .	81·5	76·0	74·5	26	,,
Methyl alcohol . .	Acetone	65·5	56·6	55·95	13·5	P.
Ethyl alcohol . . .	Carbon disulphide .	77·8	45·7	42·0	9	Ry.
Isopropyl alcohol. .	Carbon disulphide .	81·5	45·7	44·0	9	,,
Acetone	Carbon disulphide .	56·4	46·2	39·25	34	Ry., Y.
Acetone	Methyl acetate . .	56·0	56·0	55·5	...	Ry.
Ethyl iodide . . .	Acetone	72·0	56·0	55·5	40	Ry., Y.
Methyl acetate . .	Carbon disulphide .	56·0	45·6	39·5	29	,,
Ethyl acetate . . .	Ethyl iodide . . .	76·0	72·0	70·0	22	,,
Ethyl acetate . . .	Carbon tetrachloride	77·15	76·75	74·8	...	Z., Y.
Ethyl acetate . . .	Carbon disulphide .	76·0	45·6	46·0(?)	8 (?)	Ry.
Carbon disulphide .	Ethyl bromide . .	45·6	38·0	37·5	32	,,
Carbon disulphide .	Methylal.	46·2	42·05	37·25	...	Z., Y.
Benzene	n-Hexane	80·2	68·95	...	...	Y. & J.
Benzene	Carbon tetrachloride	80·2	76·75	...	...	Y. & F.; L., Z.
Carbon disulphide .	Ethyl ether. . . .	46·2	34·6	34·5	...	Gu., Y.

* 746·2 mm.

TABLE 17.

Minimum Boiling Points—Ternary mixtures.

Substances in mixture.	Boiling points.		Percentage composition.	Substances in mixture.	Boiling points.		Percentage composition.
	Components.	Mixture.			Components.	Mixture.	
A. Water	100°		7·4	A. Water.	100·0°		8·1
B. Benzene	80·2	64·85°	74·1	B. Tert.butyl alcohol	82·55	67·3°	21·4
C. Ethyl alcohol . .	78·3		18·5	C. Benzene	80·2		70·5
A. Water	100·0		7·5	A. Water.	100·0		8·6
B. Isopropyl alcohol.	82·45	66·5	18·7	B. n-Propyl alcohol .	97·2	68·5	9·0
C. Benzene	80·2		73·8	C. Benzene	80·2		82·4
A. Water.	100		...	A. Water	100·0		...
B. Ethyl alcohol . .	78·3	56·6	...	B. n-Propyl alcohol .	97·2	59·95	...
C. n-Hexane . . .	68·95		...	C. n-Hexane . . .	68·95		...

TABLE 18.

Maximum Boiling Points—Binary mixtures.

Substances in mixture.		Boiling points.			Percentage of A by weight in mixture.	Observer.
A.	B.	A.	B.	Mixture.		
Water	Nitric acid	100·0°	86·0°	120·5°	32	R.
Water	Hydrochloric acid	100·0	{about −80°	110	79·76	,,
Water	Hydrobromic acid	100·0	{about −73°	126	52·5	,,
Water	Hydriodic acid	100·0	{about −35°	127	43	,,
Water	Hydrofluoric acid	100·0	19·4	120	63	,,
Water	Formic acid	100·0	99·9	107·1	23	,,
Perchloric acid	Water	110	100·0	203	71·6	,,
Pyridine	Formic acid	117·5	100·0	149		Ga., Z.
Picoline	Acetic acid	134	118	145		,,
Propionic acid	Pyridine	140	117·5	149		,,
Methyl ether	Hydrochloric acid	−21°	{about −80°	−2	61	F.
Chloroform	Acetone	61·2	56·4	64·7	80	Ry., Y.
Chloroform	Methyl acetate	60·5	56	64·5	78	Ry.

REFERENCES.

1. Young, "Distillation," *Thorpe's Dictionary of Applied Chemistry*, Vol. I., 691.
2. Young, "The Vapour Pressures and Boiling Points of Mixed Liquids, Part I.," *Trans. Chem. Soc.*, 1902, **81**, 768.
3. Young, "The Vapour Pressures and Boiling Points of Mixed Liquids, Part III.," *ibid.*, 1903, **83**, 68.
4. Young and Fortey, "The Vapour Pressures and Boiling Points of Mixed Liquids, Part II.," *ibid.*, 1903, **83**, 45.
5. Ryland, "Liquid Mixtures of Constant Boiling Point," *Amer. Chem. Journal*, 1899, **22**, 384.
6. Zawidski, "On the Vapour Pressures of Binary Mixtures of Liquids," *Zeitschr. physik. Chem.*, 1900, **35**, 129.
7. Speyers, "Some Boiling Point Curves," *Amer. Journ. Sci.*, 1900, IV., **9**, 341.
8. Friedel, "On a Combination of Methyl Oxide and Hydrochloric Acid," *Bull. Soc. Chim.*, 1875, **24**, 160.

9. Roscoe, "On the Composition of Aqueous Acids of Constant Boiling Point," *Trans. Chem. Soc.*, 1861, **13**, 146 ; 1862, **15**, 270. "On Perchloric Acid and its Hydrates," *Proc. Roy. Soc.*, 1862, **11**, 493.
10. Kuenen and Robson, "Observations on Mixtures with Maximum or Minimum Vapour Pressure," *Phil. Mag.*, 1902, VI., **4**, 116.

REFERENCES IN TABLES 17 TO 19.

F.	= Friedel, *loc. cit.*, No. 8.
Ga.	= Gardner, *Ber.*, 1889, **23**, 1587.
G. & C.	= Goldschmidt and Constam, *Ber.*, 1883, **16**, 2976.
Gu.	= Guthrie, *Phil. Mag.*, 1884, [V.], **18**, 512.
H.	= Holley, *Journ. Amer. Chem. Soc.*, 1902, **24**, 448.
J. & Y.	= Jackson and Young, *Trans. Chem. Soc.*, 1898, **73**, 176.
K.	= Konowaloff, *Wied. Ann.*, 1881, **14**, 34.
L.	= Linebarger, *Journ. Amer. Chem. Soc.*, 1895, **17**, 615 and 690.
Ne.	= Nernst, *Zeit. physik. Chem.*, 1891, **8**, 129.
N.&W.	= Noyes and Warfel, *Journ. Amer. Chem. Soc.*, 1901, **23**, 463.
P.	= Pettitt, *Journ. phys. Chem.*, 1899, **3**, 349.
R.	= Roscoe, *loc. cit.*, No. 9.
Ry.	= Ryland, *loc. cit.*, No. 5.
Th.	= Thorpe, *Trans. Chem. Soc.*, 1879, **35**, 544.
V. & D.	= Vincent and Delachanel, *Compt. rend.*, 1880, **90**, 747.
Y. & F.	= Young and Fortey, *Trans. Chem. Soc.*, 1902, **81**, 717 and 739.
Y.	= Young, *Trans. Chem. Soc.*, 1903, **83**, 77 and data not hitherto published.
Z.	= Zawidski, *loc. cit.*, No. 6.

See also Hartman, "On the First Plait in Van der Waals's Free Energy Surface for Mixtures of Two Substances," *Journ. phys. Chem.*, 1901, **5**, 425.

CHAPTER V

COMPOSITION OF LIQUID AND VAPOUR PHASES.   EXPERIMENTAL
DETERMINATIONS

Evaporation into Vacuous Space.—When two volatile
liquids—miscible, partially miscible or non-miscible—are
placed together in a vacuous space, such as that over the
mercury in a barometer tube, evaporation takes place, and,
as a rule, the composition of the residual liquid differs from
that of the vapour. It is only when the liquids form a
mixture of maximum or minimum vapour pressure—and
therefore of constant boiling point—and when it is this particular mixture that is introduced into the vacuous space,
that the composition of the vapour is the same as that of
the liquid. In all other cases the vapour is richer in the
more volatile of the two components into which the mixture
tends to separate when distilled, these components being
either the original substances from which the mixture was
formed, or one of these substances and a mixture of the two
which has a higher or lower boiling point than that of
either of the original constituents.

If the volume of vapour is relatively very small, the composition of the residual liquid will differ only slightly from
that of the original mixture, but if it is relatively very
large, and if the boiling points of the two components into
which the mixture tends to separate are not very close
together, the residual liquid will be much richer in the less
volatile component than the original mixture.

Methods Employed.—The difficulties attending the experimental determination of the composition of liquid and vapour are, in most cases, very considerable and, unless great care be taken, erroneous and misleading results may be obtained. The chief methods which have been employed are the following :—

1. A mixture of known composition is introduced into a suitable still; a relatively very small quantity is distilled over, and the composition of the distillate—and, in some cases, of the residue also—is determined either (a) from its specific gravity (Brown) (1), (b) from its refractive power (Lehfeldt) (2), Zawidski (3), (c) from its boiling point (Carveth) (4), or (d) by quantitative analysis; but for organic liquids the last method is not generally suitable. The distillation may be carried out either in the ordinary manner under constant pressure, or at constant temperature.

2. A known volume of air is passed through the mixture at constant temperature (5); the total amount of evaporation is ascertained from the loss of weight of the liquid, and the weight of one component in the vapour is determined by quantitative analysis.

3. The relation between the boiling points and the composition of mixtures of the two liquids is first determined under constant pressure, and the curve constructed. A distillation is then carried out in such a manner that the boiling points of the liquid and of the distillate can be simultaneously determined (4). The composition of the mixture in the still and of the distillate at different stages of the distillation is then ascertained from the curve.

4. A distillation is carried out with a still-head kept at a constant temperature (6). The composition of the distillate is determined in the usual manner; that of the mixture distilled is ascertained from the temperature of the still-head, the boiling-point-composition curve having been previously constructed.

First Method.—Distillations under constant pressure have been carried out by Duclaux (7) and, with great care, by F. D. Brown (1).

Brown's Apparatus.—The apparatus employed by Brown is shown in Fig. 20. " It consists of a copper vessel, s, shaped like an ordinary tin can, but provided with a long neck *a*. This neck and the upper portion of the vessel are covered

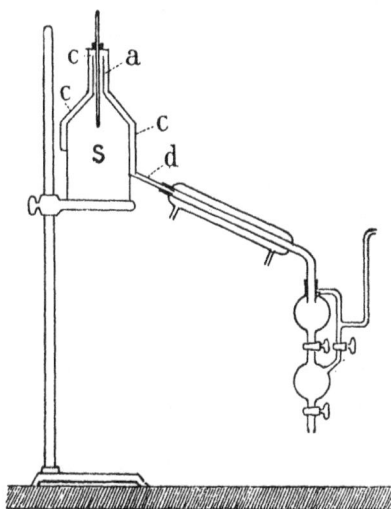

Fig. 20.—Brown's apparatus.

with a copper jacket, *c c c*, which communicates with the inner vessel by means of some small holes round the upper part of *a*. This outer jacket is terminated below by a strip of copper placed obliquely to the axis of the vessel, and at its lower portion is fitted with a narrow tube, *d*, which serves to connect the still with the condenser. The vapour rising from the liquid in the vessel s, passes through the holes at *a*, and then descending, passes out at *d*. The vapour as it rises is thus kept warm, and none of it is condensed until it has entered the outer jacket. Here a slight condensation is of no influence, as both vapour and liquid pass together into the receiver. The inclination of the bottom of the jacket serves to prevent the accumulation of any liquid at that part."

After heat has been applied, but before ebullition commences, a good deal of evaporation takes place and the mixture of warm air and vapour passes into the condenser, where most of the vapour is condensed. Brown, therefore, only made use of the data obtained from the first fraction of the distillate to ascertain the composition of the residue in the still at the moment of change from the first to the second fraction.

The form of receiver shown in Fig. 20 was used in order to avoid evaporation and consequent change in composition of the fractions, and also to allow of the distillation being carried on under reduced pressure. The quantity of liquid placed in the still was usually about 900 or 1,000 grams, and in each experiment about one-fourth of the total amount was distilled over, and was collected in four fractions. The composition of the residual liquid at the end of the distillation was found in every case to agree satisfactorily with that calculated from the composition of the original mixture and of the fractions collected.

This apparatus was employed for mixtures of carbon disulphide and carbon tetrachloride, a preliminary series of determinations of the specific gravity of mixtures of these substances having been made, in order that the composition of any mixture might subsequently be ascertained from its specific gravity. A similar series of distillations was afterwards carried out under reduced pressure (about 430 mm.), the apparatus being connected with a large air reservoir and with a pump and gauge. In the earlier experiments with mixtures of carbon disulphide and benzene somewhat less satisfactory forms of still were employed.

Lehfeldt's Apparatus.—Lehfeldt's apparatus (2) for distillation at constant temperature is shown in Fig. 21. The still, which is in the form of a large test tube, is provided with a cork perforated with two holes, through one of which passes a thermometer, the bulb of which is covered with a little cotton-wool, which dips just below the surface of the liquid. Through the other hole passes the delivery

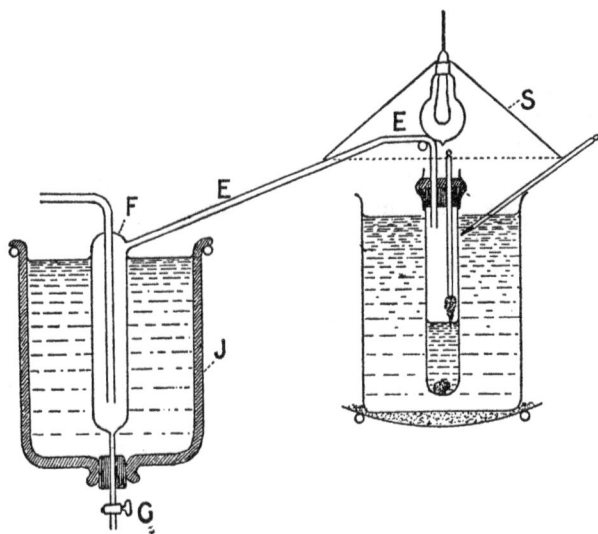

Fig. 21.—Lehfeldt's apparatus.

tube, E, E, connected with the condenser F, which is provided with a tap, G, for drawing off the distillate, and a tube connected with a pressure gauge, air reservoir and air pump. The bell-jar, J, contains either cold water or a freezing mixture. The still is heated by water in a large beaker placed on a sand bath; the water is constantly stirred and its temperature is registered by a thermometer. In order to prevent back condensation in the vertical part of the delivery tube, an incandescent electric lamp, with the ordinary conical shade, is lowered as close as possible to the water bath, and a cloth is hung round the whole; the top of the apparatus is thus kept at least as hot as the bath.

The quantities of material employed by Lehfeldt were small; about 30 c.c. of the mixture to be investigated were placed in the still and three fractions of about 1 c.c. each were usually collected and examined separately by means of a Pulfrich refractometer.

Steady ebullition was ensured by placing a piece of pumice stone, weighted with copper wire, in the still, and the pressure was adjusted from time to time to keep the boiling point as nearly constant as possible.

Preliminary determinations of the refractive powers of mixtures of the liquids investigated were made.

Zawidski's Apparatus.—Zawidski (3) employed an apparatus which is similar in principle, but more elaborate than Lehfeldt's; it is shown in Fig. 22. The still, A, of about 200 c.c. capacity, and containing in each experiment from 100 to 120 c.c. of liquid, is heated by a water bath, G, provided with a stirrer and thermostat. Back condensation is prevented by coiling copper wire round the upper part of the delivery tube, H, and heating this with a small flame. Steady ebullition is brought about by means of a fine piece of platinum wire, P, 0·04 mm. diameter, near the bottom of the still, connected to two thicker platinum leads, which pass through the side tube opposite the delivery tube and are connected with a battery of three or four accumulators. The fine wire is heated by the current of about 0·4 ampere, and a steady stream of bubbles is thus produced. This method is also recommended by Bigelow (8). The receiver, B, is of the same form as Lehfeldt's, except that the tube below, instead of being provided with a stopcock, is bent, as shown in the figure, and is connected with a second small receiver, C, into which, by diminishing the pressure in the reservoir, F, the first small portion of distillate, before the temperature and pressure have become constant, is carried over. The distillate required for examination (about 1 c.c.) is then collected in B and, after admission of air, C is removed, and, by slightly raising the pressure, the distillate is forced out of B into a little test tube.

The arrangement of the manometer, the pump and the two air reservoirs, D and F, is shown in Fig. 22.   By

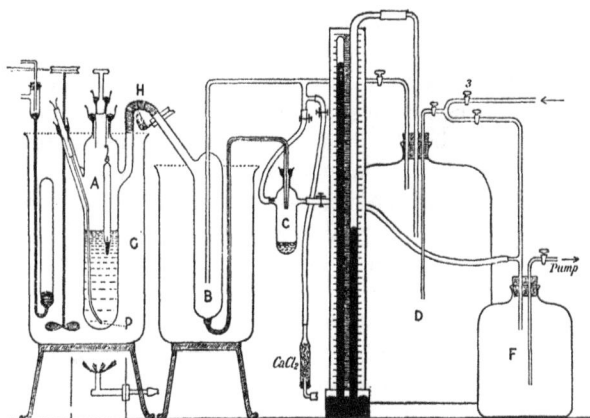

Fig. 22.—Zawidski's apparatus.

means of the various stopcocks, the pressure in F can be lowered a little below that in D, and air can be admitted into the apparatus, either through the stopcock 3 into D, or through the calcium chloride tube into C or B. A series of determinations was carried out in the following manner:— About 100 to 120 c.c. of one of the two substances was placed in A, and the pressure under which it boiled at the required temperature, t, was ascertained. A small quantity of the second liquid was then introduced into A, and the first distillation at the same temperature, t, carried out, the pressure being again noted. At the end of the distillation about 1 c.c. of the residual liquid in A was removed and placed in a small test tube for subsequent examination. A further small quantity of the second liquid was then added, and a second distillation was carried out as before, and these operations were repeated until the mixture in A became rich in the second substance.

The series of operations was then repeated, starting with the second liquid in A and adding small quantities of the first.

Sources of Error.—The following sources of error should be noted and guarded against as far as possible:—

1. When a still is heated from below in the ordinary way, the vapour which is evolved before the temperature of the upper portion of the still has reached the boiling point of the liquid will be partially condensed, and the residual vapour will contain an excessive amount of the more volatile component. Partial fractionation will, in fact, go on, and the first portions of distillate will be too rich in the more volatile component.

2. If, while the distillation is progressing, the upper part of the still or delivery tube is exposed to the cooling action of the air, partial condensation of vapour will occur and a similar error to the first will be produced.

3. The air which is in the still before the liquid is heated will become saturated with vapour, and as it passes through the condenser, part of this vapour will be condensed. In Brown's case, as already mentioned, this premature condensation of liquid was considerable in amount. The error introduced by the incomplete condensation of the vapour carried over by the air will partially compensate that referred to under number 1, and, if the top of the apparatus is heated before the liquid in the still is boiled, may possibly more than counterbalance it.

4. If a thoroughly dehydrated liquid is exposed to the air even for a short time, especially if it is poured from one vessel to another so that a large surface of it is so exposed,

a certain amount of moisture will almost invariably be absorbed. It is not only liquids which are regarded as hygroscopic which thus absorb moisture, but even substances, like benzene or the paraffins, which are classed as non-miscible with water. As a matter of fact, it is probable that no two liquids are absolutely non-miscible, and certainly all commonly occurring liquid organic compounds can dissolve appreciable quantities of water.

When such a liquid as benzene or carbon tetrachloride, containing a minute amount of dissolved water, is heated, a mixture of minimum boiling point is first formed, and the first small portion of distillate will contain the whole or at least the greater part of the water, and may probably be turbid. Suppose now that a mixture of benzene and carbon tetrachloride is being examined and that a minute amount of water has been absorbed during the preparation of the mixture or in pouring it into the still. The first portion of distillate will then contain the greater part of this water and the specific gravity, refractive index, boiling point and other physical properties of the distillate will be appreciably altered. A considerable error may thus be introduced in the estimation of the composition of the first fraction.

Brown rejected the data derived from the first distillate, and that is probably the safest plan to adopt. Zawidski rejected the first small portion of distillate and avoided the first and second sources of error, partially at any rate, by heating the delivery tube in the manner described, but in spite of this, the first results in some of his series of experiments appear to be less accurate than the later ones. Lehfeldt guarded against the first and second sources of error by heating the whole of the upper part of his apparatus by means of an incandescent lamp, as described.

It is difficult entirely to avoid the fourth source of error, and it is probably of greater influence than is generally recognised. All that can be done is to dehydrate the liquids, to dry the apparatus as completely as possible, and to keep

the liquids as far as possible out of contact with moist air; but
in any case it is best always to reject the first small portion
of distillate.

Second Method.—Determinations of the relative composition of liquid and vapour by passing a known volume of
air through a mixture at constant temperature have been
carried out by Winkelmann (9), Linebarger (5), Gahl (10),
and others.

In Linebarger's experiments a known volume of air (from
1 to 4 litres) was passed, at the rate of about 1 litre per hour,
through the liquid mixture (40—80 grams) contained in a
Mohr's potash apparatus consisting of five small and two
large bulbs. This apparatus was completely immersed in a
suitable water bath, the temperature of which was kept
constant within 0·05°.

When one of the components of the mixture was an acid,
the amount of it in the vapour was estimated by absorption
in potash or barium hydrate, and when one component contained sulphur or a halogen, the process of analysis by
means of soda-lime was adopted. The total quantity of
liquid evaporated, rarely more than 2 grams, was ascertained
by weighing the bulbs containing the mixture before and
after the experiment.

Preliminary determinations of the vapour pressures of a
number of pure liquids were made in order to test the
accuracy of the method; the results were more satisfactory
with the less volatile liquids—such as chlorobenzene,
bromobenzene and acetic acid—than with others, and they
appear to be more accurate at low temperatures than at
higher in the two cases in which such a comparison was
made.

One pair of liquids investigated by Linebarger was benzene
and carbon tetrachloride; the results obtained appear im-
probable and they differ widely from those of Lehfeldt (2),
of Zawidski (3), and of Young and Fortey (11), which are

themselves in good agreement. It is to be feared, therefore, that for mixtures of volatile liquids the method cannot be regarded as satisfactory, and even in other cases the results must be received with caution.

The tendency—so far as pure liquids are concerned—is apparently for the vapour pressures to be too low, and it may be that the air was not completely saturated with vapour, or that partial condensation occurred before the mixed air and vapour was analysed.

It is possible that the apparatus employed by Gahl (Fig. 23) would give better results.

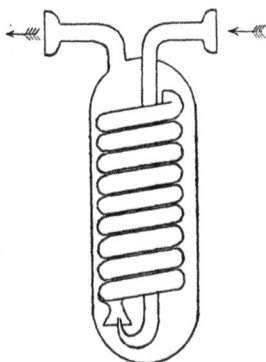

FIG. 23.—Gahl's apparatus.

Third Method.—This method was devised and employed by Carveth (4).

The boiling points of a series of mixtures of the two liquids under investigation are first determined, a round-bottomed litre-flask, provided with a reflux condenser and a thermometer, being used. The quantity of liquid should never be less than 400 c.c. After the completion of the boiling point determinations a third hole is made in the cork of the litre-flask and through it is passed a glass tube A B (Fig. 24). The length of this is about 18 cm., the diameter 20 mm., and the capacity of the bulb up to the side opening, D, about 8 c.c. Through the drawn-out portion of the bulb B is sealed a rather heavy platinum wire G to promote ebullition of liquid, and this may be further facilitated by placing tetrahedra of silver, copper or platinum in the bulb. The glass tube E serves to catch the condensed liquid from the reflux condenser and to deliver it through a small opening at the end of the tube into B. The thermometer in A B is so placed that no part of its

G

bulb is as high as the opening D. The depth to which the bulb B is immersed in the liquid in the outer vessel depends on the difference between the boiling points of the solution and of the condensed vapour; if this difference is less than 1° the liquid in B cannot be made to boil vigorously; but, as Carveth points out, steady ebullition might be ensured by means of electric heating (p. 76). In an actual experiment the liquid formed by condensation of vapour in the reflux condenser is returned (wholly or in part) to the bulb B, where it boils vigorously, and the boiling point of this distillate is read on the thermometer in A B, that of the residue being read at the same time on the thermometer in the outer bulb. The composition corresponding to each of these two temperatures is then ascertained from the boiling point-composition curve, and the composition of liquid and vapour is stated to be thus determined.

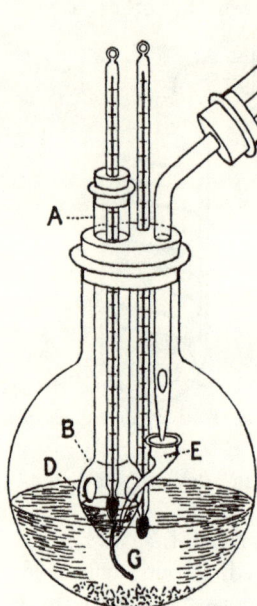

FIG. 24.—Carveth's apparatus.

It was found that the boiling point of the liquid in AB, after becoming constant, rarely varied 0·05° in the course of an hour, and it is therefore clear that a definite state of equilibrium is attained.

It is assumed that the liquid which collects in B has the same composition as the vapour evolved from the liquid in the outer flask; but, as Carveth admits, the condensed liquid, while passing from the condenser to B, is exposed to a higher temperature than its boiling point, and it seems difficult to believe that no vaporisation takes place during the descent. But, if vaporisation does occur, an excess of the more volatile component must be removed and the boiling point must therefore be raised.

On the other hand, as the liquid in B is richer in the more volatile component than that in the outer flask, the vapour from it must also be richer in that component. The vapours mix together in the outer flask and the mixture passes into the condenser. The vapour as a whole must therefore be somewhat richer in the more volatile component than that evolved from the liquid in the outer flask, and the liquid formed in the condenser must also be too rich in the more volatile component. The two errors compensate each other to some extent, but the assumption that this compensation is perfect seems to be rather a bold one.

Comparison of Brown's and Carveth's Results.—As regards the preliminary determination of the boiling points of mixtures of the two substances, it may be pointed out that, in the apparatus used, the volume of vapour—with the minimum amount of liquid, 400 c.c.—is half as large again as that of the liquid. With two substances of widely different volatility the residual liquid would certainly be very appreciably poorer in the more volatile component than the original mixture placed in the flask and the observed boiling point would therefore be too high. In the case of mixtures of carbon disulphide and benzene Carveth's boiling point-composition curve is distinctly flatter than Brown's, in other words, Carveth's boiling points are higher (allowing for a want of agreement between the observed boiling points of pure carbon disulphide).

Carveth made three series of determinations with mixtures of these two liquids. In the first, the boiling points of liquid and condensed vapour were determined for a series of mixtures of known composition, but an earlier form of apparatus was used in which there was more probability of re-evaporation of condensed liquid taking place.

In the second series the mixture was actually distilled, condensation in the ascending part of the delivery tube being prevented by means of an electric coil surrounded by asbestos. The boiling points of the distillate were separately determined, those of the mixture before and after the collection of each fraction being noted, and the mean being taken as the boiling point of the liquid corresponding to that fraction. Carveth admits that there may have been a little loss of carbon disulphide by evaporation, which would make the boiling points of the distillates too high.

In the third series, the boiling points of liquid and condensed vapour were determined in the improved form of apparatus.

The relation between the composition and the boiling points of mixtures was determined only in series I.

In series II and III the differences between the boiling points of residue and distillate agree very fairly well together, except for two observations in II, where there are apparently two mistakes of a whole degree each; but these differences do not agree at all well with those derived from the first series.

Calculating the molecular percentages of carbon disulphide in the vapour corresponding to definite percentages in the liquid, the results deduced from series II and III are considerably higher than from series I, but are much lower than those found by Brown by Method I, as will be seen from Table 19.

It is evident that either Brown's or Carveth's results must be inaccurate; in favour of Carveth's may be mentioned the fair agreement between the results of the second and third series (which are not given separately in Table 19).

TABLE 19.

Molecules per cent. of carbon disulphide in liquid.	Boiling points				Molecules per cent. of carbon disulphide in vapour.		
	Of liquid.		Liquid –;distillate.				
	Brown.[1]	Carveth.	Carveth.		Brown.	Carveth.	
	760 mm.	741 mm.	I.	II. & III.		I.	II. & III.
0	80·2°	79·70°	0°	0°	0	0	0
5	76·2	76·00	4·65	5·65	18·0	12·55	14·4
10	72·5	72·80	6·95	8·28	30·6	23·5	26·5
20	66·6	67·50	7·95	8·90	48·7	39·6	42·5
30	62·2	63·05	7·78	8·45	61·2	53·6	56·1
40	58·8	59·43	6·70	7·15	70·7	63·9	65·55
50	55·7	56·35	5·12	5·63	77·7	70·4	72·95
60	53·0	53·62	3·75	4·05	82·9	77·0	78·5
70	50·9	51·34	2·63	2·79	87·7	82·95	83·9
80	49·2	49·28	1·65	1·73	91·8	88·7	89·25
90	47·8	47·40	0·80	0·83	95·9	94·55	94·9
100	46·6	45·70	0	0	100·0	100·0	100·0

Brown's experiments with carbon disulphide and benzene were not carried out with the improved form of apparatus, which he finally employed for mixtures of carbon disulphide and carbon tetrachloride, and they are therefore probably less accurate than those with the second pair of liquids; on the other hand his later experiments with a still-head of constant temperature (Method IV) agreed fairly well in both cases with those given by Method I; indeed the results obtained by Method IV differ from Carveth's somewhat more widely than those given in Table 19.

It seems desirable that very careful determinations of the relative composition of liquid and vapour should be made with the same pair of substances by all the different methods.

Fourth Method.—In the course of his experiments with a still-head kept at a constant temperature in a bath of liquid (p. 181), Brown (6) observed that the composition of the distillate was independent of that of the mixture in the still, and depended only on the temperature of the still-head. Mixtures of carbon disulphide with carbon tetra-chloride and with benzene were specially examined and, for these two pairs of liquids, Brown had previously determined the relations between the composition of liquid and vapour, and also the relations between boiling point and composition of liquid by Method I.   From the curves representing these relations he read

1. The composition of the mixtures which would evolve vapour of the same composition as the distillates obtained with the still-head of constant temperature ;

2. The boiling points of these mixtures.

He found that these boiling points, read from the curves, agreed closely with the temperatures of the still-head. The actual numbers are given in Table 20.

In the last four determinations with carbon disulphide and carbon tetrachloride, the distillation was continued until not more than an occasional drop of distillate fell ; this may probably account for the larger differences between the two temperatures.

The agreement is sufficiently close with both pairs of liquids to allow of the statement being made that the temperature of the still-head is, approximately at any rate, the same as the boiling point of a mixture which, when distilled as in Method I, would give a distillate of the same composition as that actually collected.

Special experiments further to test the truth of this statement were made with mixtures of ethyl alcohol and water, but in this case the comparison was made by determining

1. The composition of the distillates when the still-head was kept at 81·8° and 86·5° respectively,

2. The composition of the distillate from mixtures which boiled at the same two temperatures. The results obtained are shown in Table 21.

TABLE 20.

I.	II.	III.	IV.	V.
Temperature of still-head $= t_1$.	Percentage of A in distillate.	Percentage of A in mixture evolving vapour of the composition given in Column II.	Boiling point of mixture referred to in Column III. $= t_2$.	Δ $t_1 - t_2$

A = Carbon tetrachloride ; B = Carbon disulphide.

48·1°	6·7	14·1	48·4°	−0·3°
51·5	15·7	32·9	51·8	−0·3
59·4	35·1	61·3	59·3	+0·1
63·7	48·5	73·2	63·6	+0·1
66·5	57·6	80·0	66·4	+0·1
72·1	78·5	91·6	71·8	+0·3
69·6	68·2	86·7	68·9	+0·7
70·0	69·6	87·4	69·0	+1·0
56·0	26·3	50·7	55·8	+0·2
56·5	27·6	52·2	55·9	+0·6

A = Benzene ; B = Carbon disulphide.

48·6°	6·7	16·3	48·9°	−0·3°
52·1	15·1	36·0	52·5	−0·4
56·7	24·3	53·1	57·3	−0·6
63·6	43·0	73·7	63·6	0·0
70·2	62·3	86·7	70·1	+0·1

TABLE 21.

Temperature of still-head and boiling point of mixture $= t°$.	Percentage of alcohol in distillate.	
	With still-head at $t°$.	From mixture boiling at $t°$.
81·8°	76·46	76·32
86·5	65·86	65·88

It will be seen that the agreement is very satisfactory, and it is evident that this method may be used to determine the relation between the composition of liquid and of vapour if the boiling point-composition curve has previously been constructed. Further experiments are desirable to test its applicability and the accuracy attainable.

The method would certainly possess the following advantages :—

1. The first part of the distillate might be altogether rejected, and the errors already referred to thus avoided ;

2. It would not be necessary to determine the composition of the mixture in the still.

On the other hand, this method, like Carveth's, would not be suitable for mixtures the boiling points of which vary very slightly with change of composition, such, for example, as mixtures of normal hexane and benzene containing, say, from 1 to 20 per cent. of benzene.

REFERENCES.

1. F. D. Brown, "Theory of Fractional Distillation," *Trans. Chem. Soc.*, 1879, **35**, 547 ; "On the Distillation of Mixtures of Carbon Disulphide and Carbon Tetrachloride," *ibid.*, 1881, **39**, 304.
2. Lehfeldt, "Properties of Liquid Mixtures, Part II.," *Phil. Mag.*, 1898, [V.], **46**, 42.
3. Zawidski, "On the Vapour Pressures of Binary Mixtures of Liquids," *Zeit. physik. Chem.*, 1900, **35**, 129.
4. Carveth, "The Composition of Mixed Vapours," *Journ. Phys Chem.*, 1899, **3**, 193.
5. Linebarger, "The Vapour Tensions of Mixtures of Volatile Liquids," *Journ. Amer. Chem. Soc.*, 1895, **17**, 615.
6. Brown, "The Comparative Value of different Methods of Fractional Distillation," *Trans. Chem. Soc.*, 1880, **37**, 49 ; "Fractional Distillation with a Still-head of Uniform Temperature," *ibid.*, 1881, **39**, 517.
7. Duclaux, "Tension of the Vapour given off by a Mixture of Two Liquids," *Ann. Chim. Phys.*, 1878, [5], **14**, 305.
8. Bigelow, "A Simplification of Beckmann's Boiling Point Apparatus," *Amer. Chem. Journ.*, 1899, 280.

9. Winkelmann, "On the Composition of the Vapour from Mixtures of Liquids," *Wied. Ann.*, 1890, **39**, 1.
10. Gahl, "Studies on the Theory of Vapour Pressure," *Zeit. physik. Chem.*, 1900, **33**, 179.
11. Young and Fortey, "The Vapour Pressures and Boiling Points of Mixed Liquids, Part II.," *Trans. Chem. Soc.*, 1903, **83**, 45.

CHAPTER VI

COMPOSITION OF LIQUID AND VAPOUR PHASES, CONSIDERED THEORETICALLY

Simplest Cases.—It has been seen that when two liquids are placed together in a closed vacuous space, the vapour pressure and the boiling point can only be accurately calculated from the vapour pressures of the components if (*a*) the liquids are non-miscible, or (*b*) they are miscible in all proportions and show no change of temperature or volume when mixed together, this being generally the case when the substances are chemically closely related. A similar statement may be made with regard to the composition of the vapour evolved from the two liquids.

NON-MISCIBLE LIQUIDS

The simplest case is that in which the two substances are non-miscible, for the composition of the vapour—like the vapour pressure and the boiling point—is independent of the relative quantities of the components, provided that they are both present in sufficient quantity and that evaporation can take place freely; the composition of the vapour can be calculated if the vapour pressures and vapour densities of the non-miscible liquids are known.

Calling the vapour densities D_A and D_B, and the vapour pressures at the temperature t, P_A and P_B, we shall have, in a litre of the mixed vapour, 1 litre of A at $t°$ and P_A mm. pressure and 1 litre of B at $t°$ and P_B mm. The masses of vapour will therefore be $\dfrac{0·0899 \times D_A \times 273 \times P_A}{(273 + t) \times 760}$ and $\dfrac{0·0899 \times D_B \times 273 \times P_B}{(273 + t) \times 760}$ respectively, and the relative masses will be $\dfrac{D_A P_A}{D_B P_B}$ [Naumann (1), Brown (2)].

Chlorobenzene and Water.—As an example we may again consider the case of chlorobenzene and water, the vapour pressures of which at 90° to 92° are given below.

TABLE 22.

Temperature.	Vapour Pressures in mm.		
	Chlorobenzene.	Water.	Total.
90°	208·35	525·45	733·8
91	215·8	545·8	761·6
92	223·45	566·75	790·2

The vapour density of chlorobenzene $= 56·2$ and of water $= 9$.

At 90° the relative masses of vapour $\dfrac{x_A}{x_B} = \dfrac{56·2 \times 208·35}{9 \times 525·45} = 2·48.$

At 91° ,, ,, ,, $\dfrac{x_A}{x_B} = \dfrac{56·2 \times 215·8}{9 \times 545·8} = 2·47.$

At 92° ,, ,, ,, $\dfrac{x_A}{x_B} = \dfrac{56·2 \times 223·45}{9 \times 566·75} = 2·46.$

and the percentages of chlorobenzene by weight will therefore be 71·3, 71·2 and 71·1, respectively, at the three temperatures.

In the actual experiment (p. 52), in which 80 grams of water and 110 grams of chlorobenzene were distilled together under a barometric pressure of 740·2 mm. until

about 3 grams of chlorobenzene and 40 of water remained
in the flask, the distillate was collected in five fractions,
which were found to contain the following percentages of
chlorobenzene:

$$
\begin{array}{ll}
1 & 72 \cdot 5 \\
2 & 71 \cdot 5 \\
3 & 72 \cdot 0 \\
4 & 70 \cdot 4 \\
5 & 71 \cdot 7 \\
\end{array}
$$

Mean 71·6   Calculated 71·2

Water is apt to adhere in drops to the walls of a glass
tube, but chlorobenzene flows much more freely; the first
fraction is therefore certain to contain too little water, and
it would be fairer to reject it and to take the mean from
the other four fractions. This would give 71·4 per cent. of
chlorobenzene, which agrees better with the calculated
value.

The vapour pressure, boiling point, and vapour composition
can be calculated in a similar manner when more than two
non-miscible liquids are present.

PARTIALLY MISCIBLE LIQUIDS

If the two liquids are miscible within limits, the observed
vapour pressure, boiling point and vapour composition will
differ from those calculated in the manner described, but
the difference will be small if the miscibility is only slight.

Aniline and Water.(3)—A mixture of 50 c.c. of aniline and
100 c.c. of water was distilled under a pressure of 746·4 mm.
The percentage of aniline, calculated on the assumption that
the liquids are non-miscible, would be 23·6, while the values
observed were 18·7, 20·1, 19·7, 20·4, 20·4, 19·1. After the
sixth fraction had been collected the aniline was in great
excess; a large quantity of water was added and the distillation continued, when the distillate contained 19·5 per

cent. of aniline. The composition of the distillate is clearly independent—within the somewhat wide limits of experimental error—of that of the mixture in the still, but the mean percentage of aniline, 19·9, is 3·7 lower than that calculated.

Isobutyl Alcohol and Water.—When the mutual solubility is greater, the difference between the observed and calculated values, as regards both temperature and vapour composition, is more marked. Thus isobutyl alcohol and water are miscible within fairly wide limits. At 0° a saturated solution of water in the alcohol contains 15·2 per cent. of water, and the solubility increases as the temperature rises. At 18° one part of alcohol dissolves in 10·5 of water, but the solubility diminishes with rise of temperature, reaching a minimum at about 52°.

From Konowaloff's data (4) for the vapour pressures of isobutyl alcohol, the boiling point of the alcohol and water, when distilled together under normal pressure, would be 85·7° if the liquids were non-miscible ; the minimum boiling point observed by Konowaloff was actually 90·0°, or 4·3° higher (Young and Fortey, 89·8°). Again, the calculated percentage of isobutyl alcohol in the vapour would be 74·5, while that corresponding to the minimum boiling point is actually 66·8, the difference being 7·7.

It is clear, then, that when two non-miscible liquids are heated together we can calculate the vapour pressure, the boiling point, and the vapour composition with accuracy from the vapour pressures of the components, but that, if the liquids are miscible within limits, the calculated values differ from those actually observed, the difference increasing with the mutual solubility of the liquids.

Infinitely Miscible Liquids

In the case of liquids which are miscible in all proportions, it is only, as has already been pointed out, when no appreciable change of volume or temperature occurs on admixture that the vapour pressures and the boiling points can be accurately calculated. The relation between the composition of the mixed liquid and that of the vapour evolved from it appears also to be a simple one only when the vapour pressure of the mixture is accurately expressed by the formula $P = \dfrac{mP_A + (100 - m)P_B}{100}$.

Formula of Wanklyn and Berthelot.—Many attempts have been made to find a general formula to represent the relation between the composition of liquid and of vapour. In 1863, Wanklyn (5) and Berthelot (6) arrived independently at the conclusion that the composition of the vapour depends (a) on that of the liquid, (b) on the vapour pressures of the pure components at the boiling point of the mixture, (c) on the vapour densities of the components.

Calling x_1 and x_2 the relative masses of the two substances in the vapour, W_1 and W_2 their relative masses in the liquid, D_1 and D_2 their vapour densities and P_1 and P_2 their vapour pressures, the formula would be

$$\frac{x_1}{x_2} = \frac{W_1 D_1 P_1}{W_2 D_2 P_2}.$$

In 1879, Thorpe (7) observed that carbon tetrachloride and methyl alcohol form a mixture of minimum boiling point, and that, for this particular mixture, when $\dfrac{x_1}{x_2} = \dfrac{W_1}{W_2}$, the factor $\dfrac{D_1 P_1}{D_2 P_2}$ is approximately equal to unity.

Brown's Formula.—The subject was investigated experimentally by F. D. Brown in 1879–1881 (8), and he found that the Wanklyn-Berthelot formula could certainly not be accepted as generally true. A better result was obtained with the formula

$$\frac{x_1}{x_2} = \frac{W_1 P_1}{W_2 P_2},$$

but the agreement between the calculated and observed results was still closer when a constant, c, was substituted for the ratio P_1/P_2.

Applicability of Formula.—Brown's formula,

$$\frac{x_1}{x_2} = c \cdot \frac{W_1}{W_2},$$

is not generally applicable to liquids which are miscible in all proportions, and it is obvious that it cannot be true for two liquids which form a mixture of constant boiling point. The experimental evidence, however, which has so far been obtained, points to the conclusion that Brown's law is true for those infinitely miscible liquids for which the relation $P = \frac{mP_A + (100 - m)P_B}{100}$ holds good.

The number of such cases so far investigated is, perhaps, too small to allow of the statement being definitely made that the law is true in such cases, but the evidence is certainly strong, for not only does Brown's formula hold good, within the limits of experimental error, for the two pairs of liquids investigated for which the relation between vapour pressure and molecular composition is represented by a straight line, but also, in other cases, it is found that the closer the approximation of the pressure-molecular composition curve to straightness, the smaller is the variation in the value of the " constant " c.

Zawidski's Results.—The strongest evidence is afforded by two pairs of liquids, ethylene and propylene dibromides and benzene and ethylene dichloride, investigated by Zawidski (9).

In the following tables are given the observed vapour pressures and those calculated from the formula

$$P = \frac{mP_A + (100 - m)P_B}{100},$$

also the observed molecular percentages of A in the vapour
and those calculated from the formula $\dfrac{x_B}{x_A} = c\,\dfrac{W_B}{W_A}$.

[In this formula x and W may, of course, represent either
masses or gram-molecules.]

TABLE 23.

A = Propylene dibromide ; B = ethylene dibromide ; $c = 1\cdot31$;
Temperature = $85\cdot05°$; P_B/P_A at $85° = 1\cdot357$.

Molecular percentage of A in liquid.	Vapour pressures.			Molecular percentage of A in vapour.		
	Observed.	Calculated.	Δ	Observed.	Calculated.	Δ
0·00	172·6	172·6	0·0	...	...	...
2·02	171·0	171·7	+0·7	1·85	1·55	-0·30
7·18	168·8	169·3	+0·5	6·06	5·60	-0·46
14·75	165·0	165·9	+0·9	12·09	11·66	-0·43
22·21	161·6	162·5	+0·9	18·22	17·89	-0·33
29·16	158·7	159·4	+0·7	23·50	23·90	+0·40
30·48	158·9	158·8	-0·1	23·96	25·08	+1·12
40·62	154·6	154·2	-0·4	34·25	34·31	+0·06
41·80	153·4	153·6	+0·2	34·51	35·41	+0·90
52·63	149·6	148·7	-0·9	45·28	45·89	+0·61
62·03	143·3	144·4	+1·1	55·35	55·50	+0·15
72·03	140·5	139·9	-0·6	65·86	66·28	+0·42
80·05	136·8	136·3	-0·5	74·94	75·39	+0·45
85·96	133·9	133·6	-0·3	82·45	82·38	-0·07
91·48	130·9	131·1	+0·2	89·50	89·13	-0·37
93·46	130·2	130·2	0·0	92·31	91·60	-0·71
96·41	128·4	128·8	+0·4	96·41	95·35	-1·06
98·24	127·3	128·0	+0·7	99·39	97·71	-1·68
100·00	127·2	127·2	0·0	...	...	...

The agreement between the observed and calculated percentages of propylene dibromide in the vapour is not altogether satisfactory, but Zawidski states that the errors of
experiment were much greater for this pair of substances
than for others, owing to the small quantity of material at
his disposal. The last two observed molecular percentages
of propylene dibromide in the vapour, 96·41, and 99·39, are
obviously too high ; the last should indeed be less than
98·24.

It will be seen later that when there is a real variation in the values of c, it is in the nature of a steady rise or fall from $A = 0$ to $A = 100$ per cent., whereas, in this case, the calculated values of c would be low at each end of the table.

The value of c, 1·31, does not differ greatly from the ratio of the vapour pressures, 1·357.

This series of experiments is of special interest because it is the only one in which the two liquids are very closely related. Zawidski, however, found that the relation $P = \dfrac{mP_A + (100 - m)P_B}{100}$ held good with considerable accuracy for mixtures of benzene and ethylene dichloride, and the data for this pair of liquids are therefore given in full.

TABLE 24.

$A =$ Ethylene dichloride; $B =$ Benzene; $c = 1\cdot134$; Temperature $= 49\cdot99°$; P_B/P_A at $50° = 1\cdot135$.

Molecular percentage of A in liquid.	Vapour pressures.			Molecular percentage of A in vapour.		
	Observed.	Calculated.	Δ	Observed.	Calculated.	Δ
0	268·0	268·0	0·0	...	...	...
7·16	265·5	265·7	+0·2	...	...	...
7·07	265·8	265·8	0·0	...	...	...
15·00	263·3	263·2	−0·1	11·52	13·47	+1·95
15·00	263·8	263·2	−0·6	12·72	13·47	+0·75
29·27	258·8	258·7	−0·1	26·38	26·73	+0·35
29·27	259·3	258·7	−0·6	27·06	26·73	−0·33
29·79	259·0	258·5	−0·5	27·22	27·23	+0·01
41·56	254·7	254·8	+0·1	38·72	38·68	−0·04
41·65	255·0	254·8	−0·2	38·90	38·63	−0·27
52·15	251·3	251·4	+0·1	49·00	49·01	+0·01
52·34	252·0	251·4	−0·6	49·42	49·20	−0·22
65·66	247·3	247·1	−0·2	62·66	62·77	+0·11
65·66	247·4	247·1	−0·3	62·61	62·77	+0·16
75·42	244·1	244·0	−0·1	72·96	73·02	+0·06
75·42	243·9	244·0	+0·1	73·07	73·02	−0·05
92·06	238·7	238·7	0·0	91·00	91·09	+0·09
91·89	238·3	238·8	+0·5	90·72	90·90	+0·18
100·00	236·2	236·2	0·0	...	...	...

With the exception of the first two percentages the
agreement is good, and in the case of these two it will
be seen that the percentage of A in the liquid is the same,
while the observed percentages of A in the vapour differ by
1·2. It seems clear that little weight can be attached to the
first two observations in this series.
For this pair of liquids the constant c has almost exactly
the same value, 1·134, as the ratio of the vapour pressures,
1·135.

Linebarger's Results.—It is unfortunate that the method
employed by Linebarger (10)—passing a known volume of
air through the mixture at constant temperature—did not
give more trustworthy results, for he examined some mixtures of fairly closely related substances. It may be well
to give the results obtained with one pair of liquids, chlorobenzene and benzene, for Linebarger found that there was
no appreciable heat change on mixing these substances in
several different proportions.

TABLE 25.

A = Chlorobenzene ; B = Benzene ; $c = 7\cdot3$; $t = 34\cdot8°$;
$P_B/P_A = 7\cdot16$ at 34·8°.

Molecular percentage of A in liquid.	Vapour pressures.			Molecular percentage of A in vapour.		
	Observed.	Calcu-lated.	Δ	Observed.	Calcu-lated.	Δ
15·18	126·3	126·4	+0·1	1·33	2·39	+1·06
29·08	107·9	109·0	+1·1	6·11	5·30	-0·81
65·06	63·6	64·0	+0·4	19·37	20·32	+0·95
79·21	47·0	46·3	-0·7	35·15	34·29	-0·86

Here the difference between c and P_B/P_A is probably
within the limits of experimental error, but the number of
observations is too small, and the errors are too large to
allow of the definite statement that c is quite constant.

H

Lehfeldt's Results.—The experiments of Lehfeldt (11) show that the vapour pressures of mixtures of toluene and carbon tetrachloride are slightly lower than those calculated from the formula $P = \dfrac{mP_A + (100 - m)P_B}{100}$, while those of Lehfeldt, of Zawidski, and of Young and Fortey show that the vapour pressures of mixtures of benzene and carbon tetrachloride are somewhat higher.

For toluene and carbon tetrachloride c appears to be a constant, but with benzene and carbon tetrachloride there is a small but distinct variation.

TABLE 26.

$A =$ Toluene; $B =$ Carbon tetrachloride; $c = 2\cdot76$; $t = 50\cdot0°$;
P_B/P_A at $50° = 3\cdot33$.

Molecular percentage of A in liquid.	Vapour pressures.			Molecular percentage of A in vapour.		
	Observed.	Calcu-lated.	Δ	Observed.	Calcu-lated.	Δ
0·00	310·2	310·2	0·0	...	...	...
9·01	288·8	290·6	+1·8	...	...	...
26·02	248·5	253·7	+5·2	...	...	...
29·2	...	...	...	12·8	13·0	+0·2
36·49	226·5	230·9	+4·4	...	...	...
48·3	...	...	...	25·8	25·3	−0·5
48·58	197·7	204·7	+7·0	...	...	...
60·30	174·8	179·2	+4·4	...	...	...
65·0	...	...	...	40·3	40·2	−0·1
75·99	140·8	145·1	+4·3	...	...	...
76·0	...	...	...	53·0	53·4	+0·4
83·7	...	...	...	65·6	65·1	−0·5
86·49	117·9	121·3	+3·4	...	...	...
92·7	...	...	...	81·9	82·1	+0·2
96·01	99·0	101·7	+2·7	...	...	...
100·00	93·0	93·0	0·0	...	...	...

In this case the difference between the values of c and P_B/P_A is much greater than in the previous ones, but the agreement between the observed and calculated percentages is good.

TABLE 27.

A = Benzene ; B = Carbon tetrachloride ; $c = 0.984 + 0.003\,m$;
$t = 50.0°$; P_B/P_A at $50° = 1.145$.

Molecular percentage of A in liquid $= m$.	Vapour pressures.			Molecular percentage of A in vapour.		
	Observed.	Calculated.	Δ	Observed.	Calculated.	Δ
0·00	310·2	310·2	**0·0**	...	...	...
17·0	...	...	...	16·5	16·5	**0·0**
34·0	...	...	...	32·3	32·2	**−0·1**
38·69	302·3	295·0	**−7·3**	...	...	...
62·4	...	...	...	58·8	58·6	**−0·2**
69·56	290·0	282·9	**−7·1**	...	...	...
80·3	...	...	...	76·7	76·9	**+0·2**
84·22	281·0	277·1	**−3·9**	...	...	...
95·7	...	...	...	94·6	94·6	**0·0**
100·00	270·9	270·9	**0·0**	...	...	...

Modification of Brown's Formula.—Brown's formula
is not applicable to this pair of liquids, but by taking
$c =$ const. $(c_0) + am$ instead of $c =$ const., a good agreement
between the observed and calculated values is obtained.

Mixtures of benzene with carbon tetrachloride have also
been investigated by Zawidski with very similar results.
He finds, however, somewhat lower vapour pressures for the
pure substances, and his pressure differences are rather
larger. The formula deduced from his data would be
$c = 0.961 + 0.0036m$, and there is a very fair agreement
between the calculated and observed percentages of A in the
vapour, though not quite so good as in the above table. The
first formula for c would indicate the existence of a mixture
of minimum boiling point containing 6·7 molecules per cent.
of benzene, the second a mixture containing 10·8 molecules
per cent.

**Distillations under Constant Pressure and at Constant
Temperature.** — Of other pairs of liquids, the vapour
pressures of which do not differ very greatly from those

calculated from the formula $P = \dfrac{mP_A + (100 - m)P_B}{100}$ there
are two, carbon disulphide with carbon tetrachloride and
carbon disulphide with benzene, which have been investigated
by Brown, the latter also by Carveth, but their distillations
were carried out in the usual manner under constant pressure, not at constant temperature.

Brown observed, however, that the relation between x_1/x_2
and W_1/W_2 was the same for mixtures of carbon disulphide
with carbon tetrachloride whether the distillation was carried
out under a pressure of 432 mm., or under atmospheric
pressure. If the values of c for other substances also are
independent of, or vary only slightly with, the pressure, they
can be calculated from the results of a distillation carried
out either at constant temperature or under constant pressure.

It would not be safe to conclude, without further evidence,
that the same values of c would be obtained in other cases
by both methods, but we may at any rate assume that the
differences would not be great enough to invalidate the
general conclusions deduced from a comparison of the
behaviour of different pairs of liquids.

Indirect Evidence.—As the vapour pressures of mixtures
of carbon disulphide with carbon tetrachloride and with
benzene have not been determined at constant temperatures,
it is not possible to give tables in the same form as before,
but an estimate of the relative maximum deviations of the
vapour pressures of mixtures of these substances from those
calculated from the formula $P = \dfrac{mP_A + (100 - m)P_B}{100}$ may
be formed from the boiling point curves.

It has already been stated (p. 65) that the maximum
deviations, D_1, of the theoretical boiling point curves,
constructed on the assumption that the pressure formula
holds good, from the straight lines given by the formula
$t = \dfrac{mt_A + (100 - m)t_B}{100}$ are, for the three pairs, benzene and

carbon tetrachloride, carbon tetrachloride and carbon disulphide, benzene and carbon disulphide, $-0.02°$, $-3.34°$ and $-4.02°$ respectively, while the maximum deviations, D_3, of the actual boiling point curves from the same straight lines are $-0.91°$, $-5.8°$, and $-8.0°$ respectively. The differences, $D_3 - D_1$, are, therefore, for the three pairs of liquids, $-0.89°$, $-2.5°$, and $-4.0°$, and it follows from this that the vapour pressures of mixtures of benzene and carbon tetrachloride agree most closely with those given by the formula $P = \dfrac{mP_A + (100 - m)P_B}{100}$ while the divergence is greatest for mixtures of benzene with carbon disulphide.

In all three cases the values of c may be expressed with fair accuracy by the formula $c = c_0 + am$, and it will be seen from the table below that the order of the three pairs of liquids is the same with regard to $D_3 - D_1$ as to a.

TABLE 28.

Substances in mixture.	$D_3 - D_1$.	a
Benzene and carbon tetrachloride	-0.9	0.003
Carbon tetrachloride and carbon disulphide	-2.5	0.008
Benzene and carbon disulphide	-4.0	0.016

Cases to which Brown's Modified Formula is Inapplicable. —When, for a given pair of substances, the deviations of the vapour pressures from those calculated from the formula $P = \dfrac{mP_A + (100 - m)P_B}{100}$ are great, the values of c are not only far from constant, but their relation to m cannot be represented by such a simple formula as $c = c_0 + am$. This is well seen in the case of carbon disulphide and methylal, which form a well-defined mixture of maximum vapour pressure, and of chloroform and acetone, which form a mixture of minimum vapour pressure (Zawidski).

TABLE 29.

A = Carbon disulphide ; B = methylal ; $t = 35\cdot17°$.

Molecular percentage of A in liquid.	Vapour pressures.			$c = \dfrac{x_B}{x_A} \cdot \dfrac{W_A}{W_B}$
	Observed.	Calculated.	Δ	
0	587·7	587·7	0·0	...
10	637·3	580·4	− 56·9	0·554
20	670·0	573·1	−· 96·9	0·643
30	690·9	565·7	−125·2	0·744
40	700·7	558·4	−142·3	0·875
50	701·9	551·1	−150·8	1·041
60	696·0	543·8	−152·2	1·253
70	682·0	536·5	−145·5	1·530
80	658·9	529·1	−129·8	1·926
90	612·3	521·8	− 90·5	2·531
100	514·5	514·5	0·0	...

TABLE 30.

A = Chloroform ; B = Acetone ; $t = 35\cdot17°$.

Molecular percentage of A in liquid.	Vapour pressures.			$c = \dfrac{x_B}{x_A} \cdot \dfrac{W_A}{W_B}$
	Observed.	Calculated.	Δ	
0	344·5	344·5	0·0	...
10	324·0	339·4	+ 15·4	2·080
20	304·0	334·2	+30·2	1·878
30	284·6	329·1	+44·5	1·693
40	266·9	323·9	+57·0	1·470
50	253·9	318·8	+64·9	1·257
60	247·8	313·7	+65·9	1·034
70	251·4	308·5	+57·1	0·848
80	262·1	303·4	+41·3	0·681
90	277·1	298·2	+21·1	0·580
100	293·1	293·1	0·0	...

Influence of Molecular Association.—The variability of c is even more marked when the molecules of one of the substances are associated in the liquid state. This is the case, for example, with mixtures of benzene and ethyl alcohol, which have also been investigated by Zawidski.

At 50° the vapour pressure of benzene is higher than that of alcohol, but at 80° it is the alcohol which has the higher vapour pressure. Under atmospheric pressure $D_1 = -0.13°$; $D_3 = -11.2°$; $D_3 - D_1 = -11.1°$.

TABLE 31.

A = Ethyl alcohol ; B = Benzene ; $t = 50°$.

Molecular percentage of A in liquid.	Vapour pressures.			$c = \dfrac{x_B}{x_A} \cdot \dfrac{W_A}{W_B}$
	Observed	Calculated.	Δ	
0	270·9	270·9	**0·0**	...
8·8	350·4	266·4	**– 84·0**	0·247
12·1	369·0	264·7	**–104·3**	0·290
21·5	397·0	259·8	**–137·2**	0·493
35·5	406·0	252·7	**–153·3**	0·857
44·4	404·4	248·1	**–156·3**	1·116
56·1	397·6	242·1	**–155·5**	1·556
69·7	378·4	235·1	**–143·3**	2·184
88·6	315·0	225·4	**– 89·6**	3·863
100·0	219·5	219·5	**0·0**	...

In Fig. 25, c is plotted against m, the molecular percentage of A in the liquid, and it will be seen that the curvature is very marked.

Mathematical Investigations.—The whole question of the relations between the composition of liquid mixtures and (a) the partial pressures of the vapours of

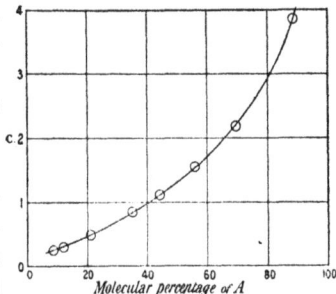

FIG. 25.—Ethyl alcohol and benzene.

the components; (b) the composition of the vapour, has been discussed mathematically by Duhem (12), Margules (13),

Lehfeldt (14), and Zawidski (9). In this connection " The Phase Rule," by Bancroft (15), may also be consulted.

Formula of Duhem and Margules.

—The formula arrived at by Duhem, and, later, by Margules, may be written :

$$\frac{d\log p_1}{d\log x} = \frac{d\log p_2}{d\log(1-x)}$$

where p_1 and p_2 are the partial pressures of the vapours of the two liquids A and B, and x and $(1-x)$ their molecular fractional amounts in the liquid mixture, taking the normal molecular weights as correct.

Lehfeldt's Formula.

—Starting from this equation, Lehfeldt adopts the formula,

$$\log t = \log K + r \log q ; \text{ or}$$
$$t = Kq^r,$$

where $t =$ the ratio of the masses of the two substances in the vapour, q the ratio in the liquid, and K and r are constants.

For r, Lehfeldt gives the equation :

$$r = \frac{\log S - \log \dfrac{\pi_A}{\pi_B}}{\log \dfrac{Bq}{A}}$$

where π_A and π_B are the vapour pressures of the pure liquids at the temperature of experiment; S is the ratio of the number of molecules of the two substances in the vapour; A and B are the normal molecular weights of the components.

Margules has pointed out that, when $r < 1$, the equation $t = Kq^r$ leads to infinite values of dp/dx when $x = 0$ or $x = 1$, which does not agree with the facts, but Lehfeldt finds that the equation holds very well for mixtures which do not contain a very small proportion of either component, provided that the molecular weights of both substances are normal in the liquid as well as in the gaseous state. For associating liquids the formula does not hold good at all.

Benzene and Carbon Tetrachloride.—As an example we may consider the case of mixtures of benzene and carbon tetrachloride, for which Lehfeldt gives the formula :

$$\log t = 0\cdot065 + 0\cdot947\log q.$$

[In Tables 32 and 33 the molecular composition is given instead of the composition by weight.]

TABLE 32.

Liquid.	Molecular percentage of A (Benzene).		
	Vapour.		
	Observed.	Calculated.	Δ
17;0	16·5	16·5	0·0
34·0	32·3	32·2	−0·1
62·4	58·8	59·2	+0·4
80·3	76·7	77·1	+0·4
95·7	94·6	94·4	−0·2

In the above equation $\log t = \log q$ when $\log q = 1\cdot2264$, that is to say, when mass of CCl_4/mass of C_6H_6 = 16·84. In other words, there would be a mixture of minimum boiling point, containing 5·6 per cent. by weight or 10·5 molecules per cent. of benzene.

Generally, if K is positive and r is less than unity, there must be a particular value of q for which $\log t = \log q$, and there must therefore be a possible mixture of constant boiling point.

Relation between Brown's and Lehfeldt's Formulæ.
—In the equation $t = Kq^r$, when $r = 1$, $t = Kq$, or Brown's

law holds good. Lehfeldt himself gives for mixtures of toluene and carbon tetrachloride the formula

$$\log t = 0\text{·}440 + 1\text{·}0 \log q,$$
$$\text{or } t = 2\text{·}755q,$$

which agrees almost exactly with that given on p. 98,

$$\frac{x_\mathrm{B}}{x_\mathrm{A}} = 2\text{·}76 \ \frac{W_\mathrm{B}}{W_\mathrm{A}}.$$

Carbon Disulphide and Methylal.—As an example of two substances, the molecular weights of both of which are presumably normal, but which form a mixture of maximum vapour pressure considerably higher than that of either component, we may take carbon disulphide and methylal, examined by Zawidski.

In Table 33 are given the observed molecular percentage of carbon disulphide and those calculated by means of the formula $\log t = 0\text{·}036 + 0\text{·}619 \log q$.

TABLE 33.

Liquid.	Vapour.			Liquid.	Vapour.		
	Observed.	Calcu-lated.	Δ		Observed.	Calcu-lated.	Δ
4·96	8·98	12·9	+3·9	60·60	54·76	54·6	−0·2
10·44	17·39	19·6	+2·2	68·03	59·21	59·5	+0·3
16·51	24·44	25·2	+0·8	73·53	62·74	63·4	+0·7
27·19	34·39	33·3	−1·1	79·27	66·76	67·8	+1·0
34·80	39·97	38·4	−1·6	84·21	70·92	72·2	+1·3
39·04	42·63	41·1	−1·5	85·73	72·83	73·6	+0·8
45·42	46·34	45·1	−1·2	91·30	80·00	80·3	+0·3
49·42	48·52	47·6	−0·9	95·76	89·23	86·4	−2·8
53·77	50·99	50·3	−0·7				

Molecular percentages of A (Carbon disulphide).

The agreement is not very good, and there appears to be evidence of curvature.

Oxygen and Nitrogen.—Baly (16) found that, for mixtures of liquid oxygen and nitrogen, Lehfeldt's formula gave very good results, but Brown's did not.

Associating Substances.—As regards mixtures of ethyl alcohol, an associating substance, with benzene and with toluene, Lehfeldt points out that the relations between $\log t$ and $\log q$ are far from linear, and he did not attempt to find an equation for the curves.

Zawidski's Formula.—Zawidski adopts the following equations to express the relations between the partial pressures, p_1 and p_2, of the components in the mixture, the vapour pressures, P_1 and P_2, of the pure components, and the molecular fractional amounts, x and $1-x$, of the two substances in the mixture.

$$p_1 = P_1 x \cdot e^{\frac{a_2}{2}(1-x)^2 + \frac{a_3}{3}(1-x)^3}$$

$$p_2 = P_2(1-x)e^{\frac{\beta_2}{2} \cdot x^2 + \frac{\beta_3}{3} \cdot x^3},$$

and $\qquad \beta_2 = a_2 + a_3 \; ; \quad \beta_3 = -a_3,$

where a_2 and a_3 are constants, the values of which can be ascertained from the partial pressure curves ; or, by Margules's method, from the tangents to the total pressure curve at the extreme points where $x = 0$ and $x = 1$ by means of the equations :

$$\left(\frac{a_2}{2} + \frac{a_3}{3}\right)\log e = \log\left[\left(\frac{d\pi}{dx}\right)_0 + P_2\right] - \log P_1$$

$$\left(\frac{a_2}{2} + \frac{a_3}{6}\right)\log e = \log\left[P_1 - \left(\frac{d\pi}{dx}\right)_1\right] - \log P_2.$$

Relation between Brown's and Zawidski's Formulæ.
—Zawidski points out that if, in his formula, the constants a and β vanish, the equations become

$$p_1 = P_1 x \text{ and } p_2 = P_2(1-x),$$

whence $\qquad p_1/p_2 = P_1/P_2 \cdot x/(1-x),$

or $\qquad p_1/p_2 = \text{const.} \times x/(1-x) \; ;$

and since, for substances of normal molecular weight, the partial pressures are proportional to the number of molecules present, the equation, in its final form, simply expresses Brown's law, taking the relative number of molecules instead of relative masses for both liquid and vapour.

Zawidski shows that this simple relation holds for mixtures of ethylene and propylene dibromides and of benzene and ethylene dichloride, but that for the first pair of liquids, the constant, 0·758, differs somewhat widely from the ratio P_1/P_2, 0·737, though the agreement is excellent for the second pair (0·880 and 0·881).

Zawidski shows, for these two pairs of substances, not only the correctness of the relation

$$P = p_1 + p_2 = P_1 x + P_2 (1 - x),$$

which is equivalent to $P = \dfrac{m P_A + (100 - m) P_B}{100}$, but also of the formulæ

$$p_1 = P_1 x \text{ and } p_2 = P_2 (1 - x),$$

as will be seen from the diagram (Fig. 26) which is taken from his paper.

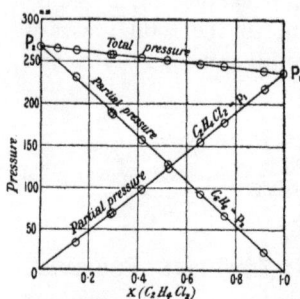

FIG. 26.—Benzene and ethylene dichloride.

In the case of other mixtures of liquids with normal molecular weight, the constants a and β were found to have finite values, and the simple formulæ $p_1 = P_1 x$ and $p_2 = P_2 (1 - x)$ were not found to be applicable; the relations between the molecular composition and the pressures, whether total or partial, are, in fact, represented by curves. Thus, even with benzene and carbon tetrachloride the curvature, though slight, is unmistakable.

Benzene and Carbon Tetrachloride.—For this pair of liquids Zawidski gives the constant a_2 and a_3, calculated by both methods ; they are—

I. From the partial pressure curve, $a_2 = 0.308$, $a_3 = 0.00733$.

II. From the tangents $(d\pi/dx)_0 = 90$, $(d\pi/dx)_1 = -4.3$, $a_2 = 0.312$, $a_3 = -0.0168$.

Now the value of the tangents $(d\pi/dx)_1$ calculated from the first constants would be -5.0, so that in either case there is a negative value for the tangent when $x = 1$. If this is correct, there must be a mixture of maximum vapour pressure containing little benzene, and there is thus additional evidence (pp. 99 and 105) that these two liquids can form such a mixture, though the difference between the maximum pressure and the vapour pressure of carbon tetrachloride is probably too small to be determined by direct experiment.

For this and other pairs of liquids of normal molecular weight, the agreement between the observed molecular composition of the vapour and that derived from the calculated partial pressures is fairly good, though there are occasionally differences amounting to 4 or 5 per cent.

Carbon Disulphide and Methylal.—In Table 34 (p. 110) are given the results for carbon disulphide and methylal, so as to compare the calculated values with those obtained by means of Lehfeldt's formula.

The agreement up to 50 molecules per cent. of carbon disulphide is excellent, but for mixtures richer in that component it is not nearly so satisfactory ; it is probable, however, that by altering the constants, better results might be obtained.

Associated Substances. — Zawidski included in his investigation some pairs of substances, of which the molecules of one component are associated in the liquid state (water), or in both the liquid and gaseous states (acetic

TABLE 34.

A = Carbon disulphide; B = Methylal.
From $(d\pi/dx)_0 = + 578$ and $(d\pi/dx)_1 = -1310$; $\alpha_2 = 2\cdot9$; $\alpha_3 = -1\cdot89$.

	Molecular percentage of Carbon disulphide.						
Liquid.	Vapour.			Liquid.	Vapour.		
	Observed.	Calculated.	Δ		Observed.	Calculated.	Δ
4·96	8·98	8·95	- 0·03	60·60	54·76	53·97	- 0·79
10·44	17·39	17·09	- 0·30	68·03	59·21	57·87	- 1·34
16·51	24·44	24·44	0·00	73·53	62·74	61·20	- 1·54
27·19	34·39	34·45	+ 0·06	79·27	66·76	65·36	- 1·40
34·80	39·97	39·97	0·00	84·21	70·92	69·88	- 1·04
39·04	42·63	42·63	0·00	85·73	72·83	71·53	- 1·30
45·42	46·34	46·25	- 0·09	91·30	80·00	79·05	- 0·95
49·42	48·52	48·36	- 0·16	95·76	89·23	88·03	- 1·20
53·77	50·99	50·56	- 0·43				

acid), and he concludes that his formula can be employed for such mixtures, provided that in calculating the values of x and $1 - x$ we take the average molecular weight of the associated liquids under the conditions of the experiment. As a rule, however, the experimental data at present available are only sufficient to afford a rough estimate of the average molecular weight of an associating substance when mixed with another liquid.

General Conclusions.—The conclusions arrived at may be stated shortly as follows :—

1. The composition of the vapour from a pair of non-miscible liquids at a given temperature may be accurately calculated from the vapour pressures and vapour densities of the components.

2. The composition of the vapour from a pair of closely related miscible liquids at a given temperature may, so far as is known, be calculated by means of Brown's formula

$x_1/x_2 = cW_1/W_2$. The value of the constant, c, does not differ greatly from the ratio of the vapour pressures of the components at the temperature of experiment, but the data at present available are insufficient to warrant the statement that it is always equal to this ratio, and it appears to be necessary to determine it experimentally.

3. As regards substances which are not closely related, Brown's formula is only applicable when the vapour pressure of any mixture is given by the formula—

$P = \dfrac{mP_A + (100 - m)P_B}{100}$, and this is probably never the case

when there is a marked volume or temperature change on mixing the pure liquids.

4. If the vapour pressures at constant temperature of mixtures of two infinitely miscible liquids—the molecular weights of which are normal—are not given by the formula

$P = \dfrac{mP_A + (100 - m)P_B}{100}$, the relation between vapour pressure

and composition must be determined experimentally at the required temperature. The composition of the vapour from any mixture may then be calculated with fair accuracy by means of the formula adopted by Zawidski (p. 107), the values of the constants a_2 and a_3 being ascertained from the pressure-composition curve.

5. When the molecules of either liquid are associated, the relation between the composition of the vapour and that of the liquid cannot be ascertained, even approximately, by means of Zawidski's formula unless the average molecular weights of the associating substance under the varying conditions of the experiment are known.

6. If, for two miscible liquids, a sufficient number of determinations of the relative composition of liquid and vapour at constant temperature have been made to allow of a curve being constructed—the molecular percentages of the vapour being mapped against those of the liquid ; x_1/x_2 against W_1/W_2 ; the logarithms of these ratios against each other,

or the partial pressures of each component separately against the molecular fractional amount of one of them— other values may be read from the curve, or the constants for an interpolation formula may be calculated.

If the vapour pressures of mixtures of the two substances differ but little from those given by the formula $P = \dfrac{mP_A + (100 - m)P_B}{100}$, a modification of Brown's formula may be used ; $x_1/x_2 = c'W_1/W_2$, where $c' = c_o + am$.

Better results are, however, generally given by Lehfeldt's formula, $\log t = K + r \log q$., and this may be used even when the observed vapour pressures differ somewhat considerably from the calculated. If, however, these differences are great, the formula of Zawidski should be employed.

Most of the investigations of the relation between the composition of liquid and of vapour have been carried out at constant temperature, but in practice a liquid is almost always distilled under constant pressure. Brown, however, whose distillations were carried out in the usual manner, found, in the case of carbon tetrachloride and carbon disulphide, that when a mixture was boiled the composition of the vapour was independent of the pressure under which ebullition took place, and, if this were generally true, a curve constructed from results obtained at constant temperature could be used to ascertain the vapour composition in a distillation under constant pressure. It is, however, to be noticed that the ratio of the vapour pressures, even of two closely related liquids, is not the same at different temperatures, and if the relation $x_1/x_2 = P_1W_1/P_2W_2$ is really true for such liquids, P_1/P_2 would be a constant for a distillation at constant temperature, but would vary slightly if the distillation were carried out under constant pressure.

At any rate Lehfeldt found that, in the case of Brown's distillations, the logarithms of the ratios of the masses of the components in the liquid and vapour phases had a linear relation, and Baly (16) obtained a similar result with the distillation of mixtures of oxygen and nitrogen under constant pressure. Lehfeldt's formula could therefore be used for interpolation in these cases.

REFERENCES.

1. Naumann, "On the Distillation of Benzene, Toluene, &c., in a Current of Steam," *Berl. Berichte*, 1877, **10**, 1421, 1819, 2015 ; "On a new Method of determining Molecular Weights," *ibid.*, **10**, 2099.

2. Brown, "Theory of Fractional Distillation," *Trans. Chem. Soc.*, 1879, **35**, 547.

3. Young, "Distillation," *Thorpe's Dictionary of Applied Chemistry*, Vol. I., 691.

4. Konowaloff, "On the Vapour Pressures of Mixtures of Liquids," *Wied. Ann.*, 1881, **14**, 34.

5. Wanklyn, "On the Distillation of Mixtures : a Contribution to the Theory of Fractional Distillation," *Proc. Roy. Soc.*, 1863, **12**, 534.

6. Berthelot, "On the Distillation of Liquid Mixtures," *Compt. rend.*, 1863, **57**, 430.

7. Thorpe, "A Contribution to the Theory of Fractional Distillation," *Trans. Chem. Soc.*, 1879, **35**, 544.

8. Brown, "The Comparative Value of different Methods of Fractional Distillation," *Trans. Chem. Soc.*, 1879, *loc. cit.* ; 1880, **37**, 49 ; "On the Distillation of Mixtures of Carbon Disulphide and Carbon Tetrachloride," 1881, **39**, 304 ; "Fractional Distillation with a Still-head of uniform Temperature," 1881, **39**, 517.

9. Zawidski, "On the Vapour Pressures of Binary Mixtures of Liquids," *Zeit. physik. Chem.*, 1900, **35**, 129.

10. Linebarger, "The Vapour Tensions of Mixtures of Volatile Liquids," *Journ. Amer. Chem. Soc.*, 1895, **17**, 615.

11. Lehfeldt, "On the properties of Liquid Mixtures," Part II., *Phil. Mag.*, 1898, [V.], **46**, 42.

12. Duhem, "On the Vapours emitted by Mixtures of Volatile Substances," *Ann. de l'École normale Sup.*, 1887, [3], **4**, 9 ; "Some Remarks on Mixtures of Volatile Substances," *ibid.*, 1889, [3], **6**, 153 ; "Solutions and Mixtures," *Trav. et Mém. de la faculté de Lille*, 1894, III. D. ; *Traité élémentaire de méchanique chimique*, 1899.

13. Margules, "On the Composition of the Saturated Vapours of Mixtures," *Sitzungsber. der Wiener Akad.*, 1895, **104**, 1243.

14. Lehfeldt, "On the Properties of Liquid Mixtures," 3 Parts, *Phil. Mag.*, 1895, [V.], **40**, 397 ; 1898, *loc. cit.* ; 1899, [V.], **47**, 284.
15. Bancroft, "The Phase Rule," 1897.
16. Baly, "On the Distillation of Liquid Air and the Composition of the Gaseous and Liquid Phases," *Phil. Mag.*, 1900, [V.], **49**, 517.

CHAPTER VII

DIRECTIONS FOR CARRYING OUT A FRACTIONAL DISTILLATION

Mixtures Separable into Two Components

It has been stated (p. 71) that when a mixture of two substances is heated, the vapour is richer than the liquid * in the more volatile of the two components into which the mixture tends to separate, whether these components are the original substances which were mixed together or a mixture of constant boiling point and one of the original substances.

If, then, we distil the mixture in the usual manner until, say, one half of the total quantity has passed over, the distillate will be richer than the residue in the more volatile component. If we were to redistil the distillate and again collect the first half, the new distillate would be still richer in the more volatile component, and by repeating the operation several times we might eventually obtain some of the more volatile component in a pure, or nearly pure, state. The amount of distillate would, however, become smaller each time, and, if a large number of distillations were required, it would be relatively very small indeed. In order to obtain a fair quantity of both components in a sufficiently pure state, systematic fractional distillation is necessary.

* Unless a mixture of constant boiling point, of the same composition as that of the original mixture, is formed, in which case the composition of the vapour would be the same as that of the liquid.

Let us suppose that we have 200 grams of a mixture of equal weights of benzene and toluene. The relation between the vapour pressures and the molecular composition of mixtures of these liquids is expressed very nearly by a straight line and it is probable that Brown's law, $\frac{x_B}{x_A} = c\,\frac{W_B}{W_A}$, is very nearly true, and that c does not differ greatly from 2·47, the mean ratio of the vapour pressures at equal temperatures between 80° and 110°. At any rate the liquid tends to separate on distillation into the original components, benzene and toluene, no mixture of constant boiling point being formed.

Collection of Distillate in "Fractions."—The mixture should be first distilled and the distillate collected in a convenient number of fractions, the receivers being changed when the boiling point reaches certain definite temperatures to be arranged beforehand. In order to trace the course of the separation as clearly as possible we will, in the first place, make the range of temperature nearly the same for most of the fractions. The boiling point of benzene is 80·2° and of toluene, 110·6°, the difference being 30·4° and we might take 10 small flasks to provide for 10 fractions, 8 with a range of 3° and 2 with a range of 3·2°, but it is better to take two fractions each for the first and last 3°. It is, as a rule, convenient to have the same number of fractions above and below the middle temperature between the two boiling points. Suitable temperature ranges for the twelve fractions are given in the second column of Table 35 (p. 116).

The first three receivers are, however, not required for the preliminary distillation.

Correction of Temperature.—Before distilling the mixture we must read the barometer, because 80·2° and 110·6° are the boiling points of benzene and toluene re-

TABLE 35.

Number of receiver.	Temperature ranges.		
	760 mm.	745 mm.	To be read on thermometer.
1	80·2— 81·2°	79·6— 80·6°	79·8— 80·8°
2	81·2— 83·2	80·6— 82·6	80·8— 82·8
3	83·2— 86·2	82·6— 85·6	82·8— 85·8
4	86·2— 89·2	85·6— 88·6	85·8— 88·8
5	89·2— 92·2	88·6— 91·6	88·8— 91·8
6	92·2— 95·4	91·6— 94·8	91·8— 95·0
7	95·4— 98·6	94·8— 97·9	95·0— 98·2
8	98·6—101·6	97·9—100·9	98·2—101·2
9	101·6—104·6	100·9—103·9	101·2—104·2
10	104·6—107·6	103·9—106·9	104·2—107·2
11	107·6—109·6	106·9—108·9	107·2—109·2
12	109·6—110·6	108·9—109·9	109·2—110·2

spectively under normal pressure, and we must find what they would be under the actual barometric pressure, and alter the temperatures accordingly. If the thermometer does not register true temperatures, the necessary corrections must be ascertained and taken into account.

Let us suppose that the height of the barometer, corrected to 0° (p. 269), is 745 mm. and that the thermometer reads 0·2° too high at 80° and 0·3° too high at 110°. Referring to p. 15 we find that the value of c $\left(= \dfrac{dt}{dp} \cdot \dfrac{1}{T} \right)$ for benzene is 0·000121 and for toluene, 0·000120. The corrections will therefore be $\Delta t = (760 - 745) (273 + 80) \cdot 0·000121 = 0·6°$ for benzene and $\Delta t = (760 - 745) (273 + 111) \cdot 0·000120 = 0·7°$ for toluene, and the boiling points under a pressure of 745 mm. will therefore be $80·2° - 0·6° = 79·6°$ and $110·6° - 0·7° = 109·9°$ respectively. The temperatures of the fractions under 745 mm. pressure are given in the third column and the actual readings on the thermometer in the fourth column.

When it is desired to separate the components of a

mixture in the purest state attainable, the above corrections must be made with the greatest possible care and it may not be sufficient to estimate the temperatures to 0·1° but to 0·05° or less ; it may also be necessary to read the barometer from time to time during the course of the distillation and and to recalculate the corrections if there is any change in the pressure.*

Rate of Distillation.—The mixture must now be slowly distilled : for laboratory purposes the drops of distillate should fall at the rate of about 1 per second, but on the large scale a much greater rate would be necessary. The slower the distillation the better is the separation and, although each distillation takes longer, time will on the whole be saved, because the same result will be attained with a smaller number of distillations. Unless otherwise stated, all the distillations recorded in this book have been carried out at the rate of 1 drop of distillate per second.

Systematic Fractional Distillation.—The results of the fractionation † of benzene and toluene with an ordinary distillation bulb, now being described, are given in Tables 36 and 37 ; the read temperatures are not stated but only the true temperatures under normal pressure.

A flask of about 270 c.c. capacity was used for the first distillation and the bulb of the thermometer was covered with a little cotton-wool to prevent super-heating (p. 12).

First Fractional Distillation.—In the first distillation the temperature rose almost at once to 86° and the first

* When a mixture of unknown composition is to be distilled, we cannot decide upon the temperature ranges of the fractions beforehand. In that case, the height of the barometer must be noted and the thermo-meter must be read when fractions of suitable bulk have been collected. The thermometer readings may be corrected afterwards.

† For the results obtained with an improved still-head, see Table 52, (p. 192).

portion of the distillate was therefore collected in the 4th receiver. On the other hand the temperature reached 110·6°, the boiling point of toluene, before the whole of the liquid had come over. The distillation was therefore stopped, the apparatus allowed to cool and the residue in the flask weighed ; This residue, amounting to 10·9 grams consisted of pure toluene and did not require to be distilled.

TABLE 36.

Number of fraction.	Temperature range.	Weight of fraction $= \Delta W$.			
		I.	II.	III.	IV.
1	80·2— 81·2°	...	...	12·95	31·55
2	81·2— 83·2	...	3·8	24·8	23·9
3	83·2— 86·2	...	33·85	22·75	16·2
4	86·2— 89·2	9·75	22·3	13·5	9·55
5	89·2— 92·2	51·8	19·65	11·8	8·0
6	92·2— 95·4	28·85	13·6	9·15	5·8
7	95·4— 98·6	21·2	12·95	7·3	5·35
8	98·6—101·6	12·8	9·05	6·75	4·65
9	101·6—104·6	11·45	8·9	6·3	3·85
10	104·6—107·6	14·15	10·8	7·95	5·85
11	107·6—109·6	13·45	9·6	8·95	7·4
12	109·6—110·6	24·9	30·75	33·05	30·5
Pure toluene	110·6	10·9	22·95	31·35	42·1
Total weight		199·25	198·2	196·6	194·7
Percentage weight of distillate below middle temperature, 95·4°		45·4	47·0	48·3	48·8

Second Fractionation.—The flask of 270 c.c. capacity was now replaced by a smaller one—about 80 c.c. and the first fraction, that in receiver No. 4, was redistilled. A small quantity came over below 83·2° and was collected in flask No. 2 ; this receiver was removed and No. 3 substituted as soon as the temperature reached 83·2°, and No. 4 was put in the place of No. 3 when the temperature had risen to 86·2°. At 89·2° the flame was removed and, after the cotton-wool on the thermometer had become dry, the second fraction from the first distillation—that in No. 5—was added to the residue in the still. On recommencing the distillation it became clear that the temperature would not reach 86·2° for a considerable time and the first portion of the distillate was therefore collected in receiver No. 3. The process was continued as before ; flask No. 4 was substituted for No. 3 at 86·2°, No. 5 for No. 4 at 89·2° and the distillation was stopped at 92·2°. The contents of receiver No. 6 were now added to the residue in the still and the distillate was collected in Nos. 4, 5 and 6 ; the remainder of the fractionation was carried out in this manner except that, after the 11th fraction from the first distillation had been placed in the still, no distillate was collected in No. 9 because the temperature rose at once to 104·6° ; so also, after the addition of the last fraction,· no distillate came over below 107·6°.

Third Fractionation.—The third fractionation was carried out in the same manner as the second, the first portion of the first distillate, and also the first portion of the distillate which came over after the contents of receiver No. 3 had been added to the residue in the still, being collected in No. 1.

Fourth Fractionation.—In the fourth fractionation, the first fraction was redistilled, the second being placed in the still when the temperature reached 81·2°. In other respects the procedure was the same as in the third fractionation.

Procedure Dependent on Number of Fractions.—If the range of temperature for each fraction had been larger, say, 5° instead of 3° for the majority of them, then on recommencing the distillation after addition of any fraction to the residue in the still, the temperature would have risen at once, or very rapidly, to the initial temperature of the fraction below; thus, on adding any fraction, say, No. 4, no distillate would have been collected in No. 2, and receiver No. 3 might have been left in position while fraction No. 4 was being placed in the still and the distillation was recommenced. This method is usually adopted, but by taking a large number of receivers there is a greater amount of separation of the components in a complete distillation and, on the whole, time is saved.

Loss of Material. —On the first distillation there was a loss of 0·75 gram of material (200·0 – 199·25), due entirely to evaporation. In the second complete distillation the total weight fell from 199·25 to 198·2, and in the subsequent distillations the loss was in nearly every case more than 1 gram. The greater loss was due partly to increased evaporation, and partly to the small amount of liquid left adhering each time to the funnel through which the fraction was poured into the still. The total loss after fourteen fractionations was 18·1 grams, a very appreciable amount.

Relative Rate of Separation of Components. —It will be observed that, in the first distillation, 10·9 grams of pure toluene were obtained, but that only 9·75 grams of distillate came over below 89·2°, a temperature 9° above the boiling point of benzene; and at the end of the fourth complete distillation the weight of pure toluene recovered was 42·1 grams, while only 31·55 grams of benzene boiling within 1 degree had been obtained. It is thus evident that the separation of the less volatile component is much easier than that of the more volatile, and this is, indeed, always found to be the case.

It will also be noticed that the weights of the middle fractions steadily diminished, while those of the lowest and highest fractions increased; this also invariably occurs when a mixture of two liquids is repeatedly distilled.*

Weight of Distillate below Middle Temperature.—The percentage weight of distillate, coming over below the middle temperature between the boiling points of the pure components, was not far below that of the benzene in the original mixture; the percentage was actually 45·4, 47·0, 48·3 and 48·8 respectively in the four distillations, while the original mixture contained 50 per cent. of benzene. Similar results are always obtained when a mixture of two liquids, which separates normally and easily into the original components, is distilled; and, as will be seen later, when an improved still-head is used, the weight of distillate below the middle temperature should, in general, be very nearly equal to that of the more volatile component in the original mixture even in the first distillation.

Alteration of Temperature Ranges.—The results of the first four complete distillations are given in Table 36, and it will be seen that if the fractionation were continued in the same manner as before, the middle fractions would soon become too small to distil, and the first fraction would never consist of pure benzene.

It is therefore preferable gradually to increase the temperature ranges of the middle fractions, and to diminish

* An account has been given by Kreis (1) of the fractional distillation of a mixture of 25 grams of benzene and 25 grams of toluene with an ordinary distillation bulb; the data for the first distillation are such as might be expected, but the results given for the second are quite impossible, for while in the first distillation 17 c.c. of liquid are stated to have been collected above 108°, in the second complete distillation the volume of the corresponding fraction is given as only 3 c.c., and it was not until the 8th fractionation that the volume again reached 17 c.c. Assuming the correctness of the data for the first distillation, the last fraction in the second should have been at least 6 times greater than that stated; it is difficult to understand how the mistake can have arisen.

those of the fractions near the boiling points of the pure substances (2). This was accordingly done in the subsequent fractionations, which were carried to their extreme limit. The results, together with those of the fourth fractionation, are given in Table 37.

Ratio of Weight of Fraction to Temperature Range.

—So long as the fractions, or most of them, are collected between equal intervals of temperature, the weights of distillate indicate clearly enough the progress of the separation, but when the temperature ranges are gradually altered this becomes less evident, and it is advisable either to plot the percentage (or total) weight of distillate against the temperature (Fig. 27), or to divide the weight of each

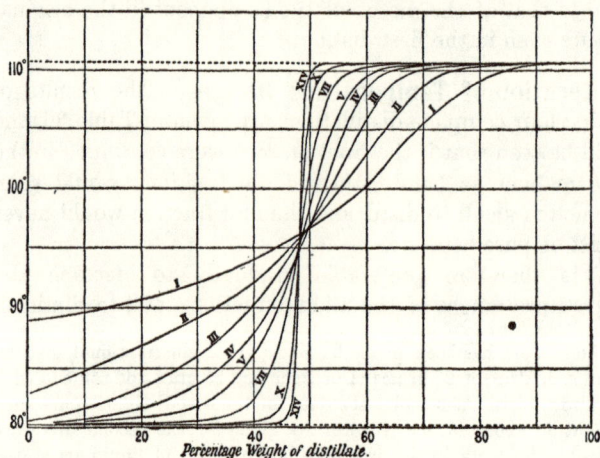

FIG. 27.—Results of fractional distillation of mixture of benzene and toluene.

fraction, Δw, by its temperature range, Δt, and to tabulate these ratios as well as the actual weights. The purer the liquid in any fraction, the higher is the ratio $\Delta w/\Delta t$, that for a pure liquid being, of course, infinitely great; and it will be seen from Table 37 that in the later fractionations, while the weights of the middle fractions diminish, and those of the lowest and highest increase much more slowly than before, the ratios $\Delta w / \Delta t$ continue to change rapidly.

Separation of Pure Benzene.—When the liquid collected below 80·45° in the seventh fractionation was distilled at the beginning of the eighth, the temperature remained at the boiling point of benzene, 80·2°, for a short time, and a considerable amount of distillate came over below 80·25°. As the range of this fraction was reduced to 0·15°, it was concluded that the first portion of the first distillate in the ninth fractionation would consist of pure benzene; it was therefore collected separately as there was no necessity to redistil it, and in each of the subsequent fractionations the first portion of the first distillate was taken to be pure benzene.

Elimination of Fractions.—In the tenth fractionation the weights of the 11th and 12th fractions had become very small, and their temperature ranges were reduced to 0·2° and 0·05° respectively. In the next fractionation, therefore, after No. 12 distillate had been added to the residue in the still, the 11th receiver was left in position until the temperature reached 110·6°; the distillation was then stopped, and the residue taken as pure toluene. In this way the number of fractions was reduced to eleven by the elimination of No. 12. In the 12th and 13th fractionations the last fraction was in each case similarly eliminated, and, in the latter, the number of fractions was further reduced by the exclusion of No. 1, and in the 14th by that of No. 2.

In the 14th fractionation, after No. 6 had been added to the residue in the still, it was found that the temperature rose at once above 81°, so that nothing was collected in No. 5, and also that when the distillation had been carried as far as possible, the temperature had not risen

TABLE 37.

No. of fraction F	Final temperature	Weight of fraction	IV. $\frac{\Delta w}{\Delta t}$	V. t	Δw	$\frac{\Delta w}{\Delta t}$	VI. t	Δw	$\frac{\Delta w}{\Delta t}$	VII. t	Δw	$\frac{\Delta w}{\Delta t}$
Pure Benzene	...	...	...	...	...	...	...	...	...	...	...	...
1	81·2°	31·55	31·55	80·9°	40·05	57·2	80·6°	42·95	107·4	80·45°	52·4	209·6
2	83·2	23·9	11·95	82·5	23·3	14·6	81·8	23·15	19·3	81·25	17·2	21·5
3	86·2	16·2	5·4	85·1	10·9	4·2	84·0	10·2	4·6	83·0	8·0	4·5
4	89·2	9·55	3·2	88·1	7·85	2·6	87·0	6·65	2·2	86·0	5·85	2·0
5	92·2	8·0	2·7	91·3	5·75	1·5	90·5	4·75	1·4	90·0	4·3	1·1
6	95·4	5·8	1·8	95·4	5·2	1·3	95·4	4·5	0·9	95·4	3·75	0·7
7	98·6	5·35	1·7	99·5	4·15	1·0	100·3	3·8	0·8	100·8	3·2	0·6
8	101·6	4·65	1·5	102·7	3·8	1·2	103·8	3·65	1·0	104·8	3·2	0·8
9	104·6	3·85	1·3	105·7	4·5	1·5	106·8	4·0	1·3	107·8	3·6	1·2
10	107·6	5·85	1·9	108·3	5·2	2·0	109·0	4·85	2·2	109·55	4·9	2·8
11	109·6	7·4	3·7	109·9	7·25	4·5	110·2	8·5	7·1	110·35	8·7	10·9
12	110·6	30·5	30·5	110·6	28·1	40·1	110·6	20·4	51·0	110·6	13·05	52·2
Pure Toluene		42·1	∞		46·8	∞		54·05	∞		61·75	∞
Total weight		194·7			192·85			191·45			189·9	
Percentage weight of distillate below middle point =			48·8		48·3			48·2			48·2	

TABLE 37 (*continued*).

F.	VIII. t	Δw	$\dfrac{\Delta w}{\Delta t}$	IX. t	Δw	$\dfrac{\Delta w}{\Delta t}$	X. t	Δw	$\dfrac{\Delta w}{\Delta t}$	XI. t	Δw	$\dfrac{\Delta w}{\Delta t}$
	...	...	...	80·2°	10·2	∞	80·2°	23·35	∞	80·2°	44·4	∞
1	80·35°	54·45	**363·0**	80·3	45·0	**450·0**	80·25	36·25	**725·0**	80·25	17·1	**342·0**
2	80·85	15·6	**31·2**	80·6	17·55	**58·5**	80·45	12·2	**61·0**	80·35	10·25	**102·5**
3	82·0	8·6	**7·5**	81·4	7·2	**9·0**	80·95	8·75	**17·5**	80·6	8·0	**32·0**
4	84·7	5·5	**2·0**	83·5	4·5	**2·1**	82·4	4·0	**2·8**	81·4	3·75	**4·7**
5	89·0	3·55	**0·8**	88·0	3·3	**0·7**	86·4	3·2	**0·8**	85·0	3·65	**1·0**
6	95·4	3·2	**0·5**	95·4	2·75	**0·4**	95·4	2·3	**0·3**	95·4	2·4	**0·2**
7	101·8	3·0	**0·5**	102·8	2·6	**0·4**	104·4	2·65	**0·3**	105·8	2·25	**0·2**
8	106·1	2·85	**0·7**	107·3	2·85	**0·6**	108·4	2·6	**0·6**	109·4	2·8	**0·8**
9	108·8	3·5	**1·3**	109·4	3·15	**1·5**	109·85	3·05	**2·1**	110·2	2·8	**3·5**
10	109·95	4·4	**3·8**	110·2	3·45	**4·3**	110·35	3·4	**6·8**	110·45	3·05	**12·2**
11	110·45	5·3	**10·6**	110·5	7·0	**23·3**	110·55	5·5	**27·5**	110·6	5·25	**35·0**
12	110·6	11·45	**76·3**	110·6	5·75	**57·5**	110·6	4·15	**83·0**			
	67·15		∞	72·1		∞	75·2		∞	79·6		∞
	188·55			187·3			186·6			185·3		
	48·2			48·3			48·3			48·3		

TABLE 37 (*continued*).

No. of fraction. F	XII.			XIII.			XIV.		
	Final temperature.	Weight of fraction.	$\frac{\Delta w}{\Delta t}$	t	Δw	$\frac{\Delta w}{\Delta t}$	t	Δw	$\frac{\Delta w}{\Delta t}$
Pure Benzene	80·2°	52·75	∞	80·2°	62·25	∞	80·2°	73·3	∞
1	80·25	10·85	217·0						
2	80·3	7·6	152·0	80·25	11·1	222·0			
3	80·4	6·35	63·5	80·3	5·45	109·0	80·25	5·45	109·0
4	80·8	5·15	12·9	80·55	2·3	9·2	80·4	3·55	23·7
5	83·4	3·6	1·4	81·4	3·4	4·0	80·9	1·05	2·1
6	95·4	2·55	0·2	95·4	3·9	0·3	91·4	3·85	0·4
7	107·4	2·35	0·2	109·4	2·8	0·2	109·8	2·95	0·3
8	110·0	2·6	1·0	110·35	2·65	2·8	110·5	2·05	2·9
9	110·4	3·15	7·9	110·6	4·4	17·6	110·6	2·1	21·0
10	110·6	4·2	21·0						
11									
12									
Pure Toluene	82·95	∞			85·0	∞		86·3	∞
Total weight	184·1				183·25			(181·9)	
Percentage weight of distillate below middle point =	48·3				48·3			...	

TABLE 37 (*continued*).

F	XV.			XVI.			XVII.			XVIII.		
	t	Δw	$\dfrac{\Delta w}{\Delta t}$	t	Δw	$\dfrac{\Delta w}{\Delta t}$	t	Δw	$\dfrac{\Delta w}{\Delta t}$	t	Δw	$\dfrac{\Delta w}{\Delta t}$
	80·2°	75·7	∞	80·2°	77·7	∞	80·2°	79·0	∞	¦80·2°	81·4	∞
3	80·3	3·85	**38·5**	80·3	2·45	**24·5**	80·3	3·5	**35·0**			
4	80·65	2·6	**7·4**	80·6	3·65	**12·2**						
5	81·85	3·25	**2·7**									
7	$\begin{Bmatrix}107·5 \\ \text{to} \\ 110·2\end{Bmatrix}$	1·95	**0·7**	$\begin{Bmatrix}110·45 \\ \text{to} \\ 110·6\end{Bmatrix}$	1·5	**10·0**						
8	110·6	2·7	**6·8**									
		87·55	∞		88·8	∞						

higher than 91·4°. The distillation was therefore stopped at this point, and the residue in the still, after cooling, was placed in a separate receiver R. Fraction No. 7 was then placed in the still, and the distillate collected in R until the temperature reached 100°, and then in No. 7. The rest of the fractionation was carried on as before.

Final Fractionations.—In the remaining distillations the fractions below and above the middle point were treated separately, the residue from the last of the lower fractions and the first portion of distillate from the first of the higher ones being collected each time in R.

Four additional operations were required for the separation of the benzene and two for that of the toluene. The total weight of pure benzene recovered by pushing the series of fractionations to their extreme limit was 81·4 grams out of 100 taken; the weight of toluene was 88·8 grams. The amount of liquid collected in R, and rejected, was 7·8 grams, and the loss by evaporation, and by transference from flasks to still, was 22·0 grams. The amount of time required for the whole operation was a little over thirty hours.

MIXTURES SEPARABLE INTO THREE COMPONENTS

We have seen that when a mixture tends to separate on distillation into two components, it is the less volatile component which is the easier to obtain in a pure state. When a mixture tends to separate into three components the least volatile of them can, as a rule, be most readily isolated and, of the other two, the more volatile is the easier to separate.

Methyl, Ethyl and Propyl Acetates.—An illustration of this is afforded by the fractionation of a mixture of 200 c.c. of methyl acetate (b.p. 57·1°), 250 c.c. of ethyl acetate (b.p. 77·15°) and 200 c.c. of propyl acetate (b.p. 101·55°) carried out with a plain vertical still-head one

metre in length. Full details of this fractionation have
been published in the *Philosophical Magazine* (3) and in
Table 38 the results of the first eight fractionations and of
12th, 16th, 20th and 24th are given.* It will be noticed that
the method adopted differed from that already described in
so far that the number of fractions was small at first and was
gradually increased. For convenience, the fractions are so
numbered that they fall into their proper places in the later
distillations. The data given are as follows :—F, the number
of the fraction ; t, the final temperature, corrected and
reduced to 760 mm. pressure, for each fraction, but in the
first four fractionations the lowest temperature (in brackets)
is that at which the distillate began to come over ; Δw,
the weight of each fraction, and the ratio $\Delta w/\Delta t$.

First and Second Fractionations.—The first fractionation, I, requires no comment ; the second, II, was carried
out in the following manner :—the first fraction from I
(No. 5) was distilled and the distillate was collected in
receiver No. 4 until the temperature rose to 63·8°, when
receiver No. 5 was substituted for it and the distillation
was continued until the temperature reached 71·0°. The
gas was then turned out and the second fraction from I
(No. 8) was added to the residue in the still. Heat was
again applied and the distillate was collected in No. 5
until the temperature rose to 71·0°, when this receiver
was replaced by No. 8, and the distillation was continued
until the temperature reached 77·1°. The third fraction
from I (No. 11) was then added to the residue and the
distillate was collected in No. 8 until the temperature
again rose to 77·1°, when No. 11 was put in its place.
At 84·4° a new receiver, No. 13, was substituted for No. 11
and the temperature was allowed to rise to 91·7°. The last
fraction from I was then added to the residue in the flask

---

* For the results obtained with an improved still-head see Table 54,
p. 194.

TABLE 38.

	I.				II.				III.		
F	t	Δw	$\dfrac{\Delta w}{\Delta t}$	F	t	Δw	$\dfrac{\Delta w}{\Delta t}$	F	t	Δw	$\dfrac{\Delta w}{\Delta t}$
A											
1											
2									(58·4°)		
3					(60·9°)			3	60·95	36·5	14·3
4	(63·8°)			4	63·8	59·7	20·6	4	63·85	77·2	26·6
5	71·3	141·4	18·9	5	71·0	134·2	18·6	5	71·0	100·6	14·0
6											
7											
8	77·8	145·2	22·3	8	77·1	98·4	16·1	8	77·15	75·4	12·4
9											
10											
11	89·2	146·9	12·8	11	84·4	67·2	9·2	11	84·4	67·7	9·3
12											
13				13	91·7	51·0	7·0	13	91·65	59·6	8·2
14 {	above 89·2	}143·3	...	14	98·4	86·1	12·7	14	98·4	46·7	6·9
15				15	101·5	63·7	20·6	15	101·45	78·3	26·1
16											
17				17	(101·55)	14·3	...	17	101·55	19·7	197·0
X											
Y											
Z								Z	...	10·0	∞
		576·8				574·6				571·7	

Table 38 (continued).

	IV.				V.				VI.		
F	t	Δw	$\dfrac{\Delta w}{\Delta t}$	F	t	Δw	$\dfrac{\Delta w}{\Delta t}$	F	t	Δw	$\dfrac{\Delta w}{\Delta t}$
(57·7°)								1	57·3°	6·3	31·5
2	58·6	17·2	19·1	2	58·2°	19·4	17·6	2	58·2	28·5	31·7
3	61·0	58·1	23·7	3	60·7	69·9	28·0	3	60·1	55·0	29·0
4	63·9	53·1	18·3	4	63·85	49·0	15·3	4	63·85	54·2	14·3
5	71·05	85·1	11·8	5	71·0	69·0	9·6	5	70·75	47·4	6·9
								7	75·7	52·0	10·4
8	77·15	66·6	10·9	8	77·15	71·1	11·6	8	77·15	33·4	22·3
11	84·3	84·7	11·8	11	83·35	85·1	13·6	11	78·4	34·8	29·0
								12	82·5	47·9	11·7
13	91·6	40·8	5·6	13	90·75	33·7	4·6	13	90·2	28·1	3·6
14	98·4	33·8	5·0	14	98·45	33·5	4·4	14	98·45	27·8	3·3
15	100·95	44·1	17·6	15	100·95	29·4	11·8	15	100·95	25·6	10·2
16	101·45	38·7	77·4	16	101·45	36·7	73·4	16	101·45	30·3	60·6
17	101·55	24·2	242·0	17	101·55	31·1	311·0	17	101·55	34·0	340·0
Z	...	20·2	∞	Z	...	34·7	∞	Z	...	50·6	∞
		566·6				562·6				555·9	

к 2

TABLE 38 (*continued*).

VII.				VIII.				XII.			
F	t	Δw	$\dfrac{\Delta w}{\Delta t}$	F	t	Δw	$\dfrac{\Delta w}{\Delta t}$	F	t	Δw	$\dfrac{\Delta w}{\Delta t}$
				A	57·1°	6·0	∞	A	57·1°	41·1	∞
1	57·3°	13·3	**66·5**	1	57·3	11·7	**58·5**	1	57·2	17·0	**170·0**
2	58·2	42·9	**47·7**	2	58·2	49·1	**54·6**	2	57·65	30·5	**67·8**
3	60·0	47·5	**26·4**	3	60·1	45·4	**23·9**	3	58·9	30·3	**24·2**
4	63·8	43·5	**11·4**	4	63·8	36·0	**9·7**	4	63·5	31·0	**6·7**
5	70·75	45·9	**6·7**	5	70·7	37·8	**5·5**	5	71·9	23·8	**2·8**
7	75·8	42·7	**8·4**	7	75·8	41·9	**8·2**	7	76·45	32·9	**7·1**
8	77·2	45·0	**32·1**	8	77·15	43·8	**32·4**	8	77·0	30·5	**55·5**
								9	77·15	27·6	**184·0**
								10	77·3	36·2	**241·3**
11	78·3	37·8	**34·4**	11	78·2	51·1	**48·7**	11	77·6	35·6	**85·3**
12	82·1	41·1	**10·8**	12	81·1	34·1	**11·8**	12	78·85	18·7	**14·9**
13	90·15	23·4	**2·9**	13	90·05	22·8	**2·5**	13	88·9	17·9	**1·8**
14	98·45	22·3	**2·7**	14	98·5	17·8	**2·1**	14	99·55	9·5	**0·9**
15	100·95	17·6	**7·0**	15	101·0	16·7	**6·7**	15	101·15	8·1	**5·1**
16	101·45	24·3	**48·6**	16	101·45	19·4	**43·3**				
17	101·55	23·7	**237·0**	17	101·55	12·6	**126·0**				
								X	...	7·8	
								Y	...	8·1	
Z	...	80·2	∞	Z	...	101·5	∞	Z	...	132·6	
		551·2				547·7				529·2	

TABLE 38 (*continued*).

	XVI.				XX.				XXIV.		
F	t	Δw	$\frac{\Delta w}{\Delta t}$	F	t	Δw	$\frac{\Delta w}{\Delta t}$	F	t	Δw	$\frac{\Delta w}{\Delta t}$
A	57·1°	79·9	∝	A	57·1°	105·5	∝	A	...	105·5	∝
1	57·2	14·6	**146·0**	1	57·15	5·9	**118·0**				
2	57·6	17·3	**43·2**	2	57·55	10·7	**26·7**	B & C	...	21·7	
3	58·6	14·6	**14·6**	3	58·55	8·9	**8·9**	D	...	10·0	
4	65·35	25·0	**3·7**	4	67·25	17·6	**2·0**	4	67·2°	6·8	**0·8**
5	73·9	15·6	**1·8**	5	75·95	16·6	**1·9**	5	75·3	15·1	**1·9**
								6	76·8	7·8	**5·2**
7	76·8	26·6	**9·2**	7	76·95	16·0	**16·0**	7	77·1	12·2	**40·7**
8	77·1	26·7	**89·0**	8	77·15	13·4	**67·0**	8	77·15	14·6	**292·2**
9	77·15	39·0	**780·0**	9	77·15	45·0	**1888·0**	9	77·15	31·9	very high
10	77·2	39·9	**798·0**	10	77·2	49·4		10	77·15	26·9	
11	77·35	24·4	**162·6**	11	77·25	13·0	**260·0**				
12	77·9	11·6	**21·1**								
U	...	6·6		S	...	8·6		Q	...	13·1	
V to Z	...	172·4		T to Z	...	192·3		R to Z	...	228·8	
		514·2				502·9				494·4	

and the distillate collected in No. 13 up to 91·7°, after which fractions were collected in No. 14 from 91·7° to 98·4° and in No. 15 from 98·4° to 101·5°, when the distillation was stopped and the residue was poured into No. 17.

Separation of Propyl Acetate.—The third fractionation III was carried out in a similar manner, a new fraction, No. 3, being collected at the beginning. As the temperature rose to 101·55°, the boiling point of propyl acetate, before the end of the last distillation, the residue was placed in a separate flask, z, and was not redistilled. The residues from subsequent fractionations up to the 10th were collected in z, but after this, as a large amount of propyl acetate had been removed, the temperature did not reach 101·55° and the residue from the 11th fractionation (b.p. above 101·45°) was placed in a new flask, y. At the end of the 12th fractionation the temperature rose only to 101·15° and the residue was placed in a third flask, x, and subsequently the residues were placed in w, v . . . . q, as shown in Table 38.

Separation of Methyl Acetate.—It was not until the 5th fractionation that the first fraction began to boil at 57·1°, the boiling point of methyl acetate, and it was not thought advisable to separate the first portion of the first distillate until the 8th fractionation. This portion and also the corresponding ones up to the 20th fractionation were collected together in a flask, A, after which the first portions were collected in B, C, D and E.

Accumulation of Ethyl Acetate in Middle Fractions. —The presence of the middle substance, ethyl acetate, is not clearly indicated until the 4th fractionation, when the value of $\Delta w/\Delta t$ for the 6th distillate (No. 11) is somewhat higher than for those above and below it ; but the gradual accumulation of the ethyl acetate in the middle fractions in subsequent fractionations is clearly shown by the rise in the

value of $\Delta w/\Delta t$ for the fractions 8 and 11 and, after the
10th fractionation, for fractions 9 and 10. The range of
temperature for Nos. 9 and 10 was gradually diminished
from 0·2° each in the 11th fractionation until no rise could
be detected; there was, indeed, no perceptible rise of
temperature during the collection of No. 11 in the 22nd
and 23rd fractionations. It was therefore certain that after
the 26th fractionation the tenth fraction was free from
propyl acetate, and that the remaining fractions B to E and
5 to 10 contained only methyl and ethyl acetates. Similarly,
it is safe to conclude that the fractions Q to Z were free
from methyl acetate. The preliminary series of fractionations was therefore completed, no fraction now containing
more than two components.

Graphical Representation of Results.—The progress
of the separation is well seen by mapping the temperatures
against the percentage weights of distillate collected, and
the curves for the first twelve fractionations are shown in
Fig. 28 (a and b). The horizontal lines at the extremities
of the later curves represent the methyl and propyl acetates
removed in the first portion of the first distillates and in the
residues respectively. The presence of ethyl acetate is
clearly indicated in the fourth curve but not in the earlier
ones.

Final Fractionations.—The fractions into which the esters
had been separated at the end of the 26th fractionation
are shown in Table 39. The total weight was 490·8 grams,
and therefore 86 grams had been lost by evaporation and
by transference from flask to still. The final separation of
methyl and ethyl acetates and of ethyl and propyl acetates
was carried out in the manner described for the later fractionations of mixtures of benzene and toluene, but it was
necessary to treat the methyl acetate with phosphorus
pentoxide to remove moisture, and the propyl acetate with

potassium carbonate to remove free acid due to slight hydrolysis. The loss was thus much greater than it would otherwise have been.

TABLE 39.

Methyl and Ethyl acetates.			Ethyl and Propyl acetates.		
Fraction.	Temperature range.	Weight.	Fraction.	Temperature range.	Weight.
A	57·1°	105·5	Q	77·15°	32·3
B	57·1 —57·15°	11·6	R	77·15— 77·2°	19·7
C	57·15—57·55	10·1	S	77·2 — 77·3	16·8
D	57·55—58·55	10·0	T	77·3 — 77·65	9·7
E	58·55—68·9	11·9	U	77·65— 79·6	10·2
5	68·9 —75·7	7·1	V	79·6 — 99·7	15·8
6	75·7 —76·9	10·4	W	99·7 —101·15	8·1
7	76·9 —77·1	15·4	X	101·15—101·45	7·8
8	77·1 —77·15	16·4	Y	101·45—101·55	8·1
9 & 10	77·15°	31·3	Z	101·55°	132·6
		229·7			261·1

The final results are given below.

TABLE 40.

	Weights.			Specific gravity at 0°/4°.*	
	Taken.	Recovered.	Percentage recovered.	Before mixing.	After fractionation.
Methyl acetate .	183	88·0	48·1	0·95932	0·95937
Ethyl acetate .	222	$\left\{\begin{matrix}56·1\\62·7\end{matrix}\right\}$ 118·8	53·5	0·92436	$\left\{\begin{matrix}0·92438\\0·92437\end{matrix}\right.$
Propyl acetate .	175	126·8	72·5	0·91016	0·91008

* Specific gravity at 0° compared with that of water at 4°.

The esters were redistilled over phosphorus pentoxide before their specific gravities were determined.

COMPLEX MIXTURES

The procedure in the case of complex mixtures is similar to that adopted when there are three components, but the number of fractions must be increased and the time required is longer.

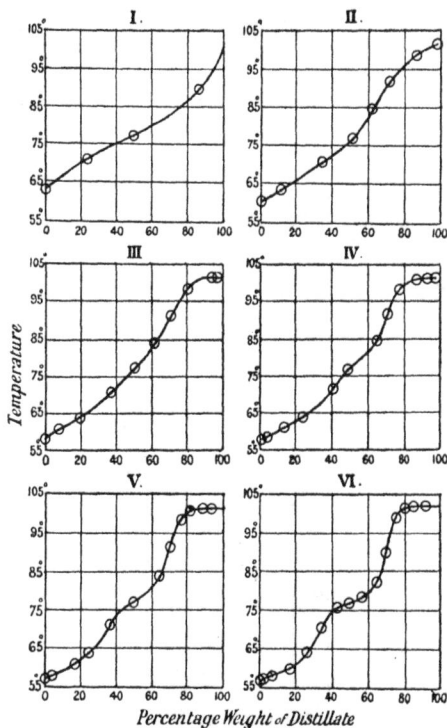

FIG. 28 (a).—Results of fractional distillation of mixture of methyl, ethyl, and propyl acetates.

If, in a complex mixture, there are two liquids boiling at temperatures not very far apart, together with others which have boiling points relatively much lower and much higher, the early fractionations do not appear, as a rule, to indicate the presence of the two liquids, but of a single substance with a boiling point between those of the two which are actually present, and it is only after some progress has been made with the fractionations that the presence of both liquids is clearly shown.

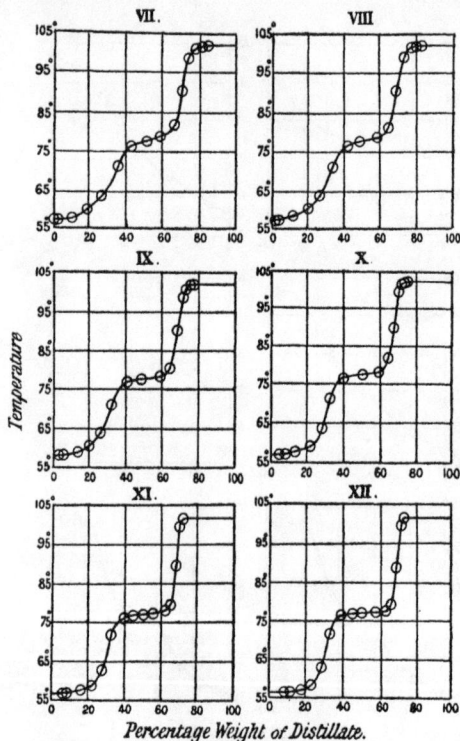

FIG. 28 (b).—Results of fractional distillation of mixture of methyl, ethyl, and propyl acetates (continued).

Separation of Pentanes from Petroleum.—A striking instance of this is afforded by the separation of iso-pentane (b.p. $27\cdot95°$) and normal pentane (b.p. $36\cdot3°$) from the light distillate from American petroleum (4).

After treatment with a mixture of concentrated nitric and sulphuric acids to remove impurities, the "petroleum ether" consists chiefly of butanes and probably a little tetra-methyl methane, all boiling below 10°, the two pentanes referred to, and hexanes and other hydrocarbons boiling higher than 60°, with very much smaller quantities of penta-methylene and trimethylethyl methane, both of which boil at about 50°; there are no substances present except the two pentanes with boiling points between about 10° and 50°.

In order to effect the separation of the pentanes it is necessary to employ a very efficient still-head (Chapter XII) but the method of arranging the fractions is the same as when an ordinary still-head is used.

Graphical Representation and Interpretation of Results.—The fractionations will be referred to more fully later on (pp. 182 and 253), but it may be well here to con-

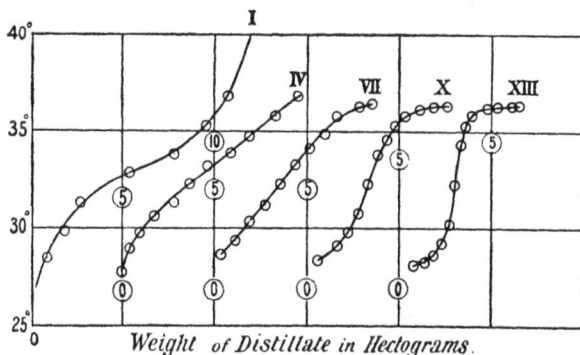

Fig. 29.—Separation of normal and isopentane from American petroleum.

sider the curves showing the results of the 1st, 4th, 7th, 10th and 13th fractionations (Fig. 29).

The first curve, I, seems to indicate (5) that a single substance boiling at about 33° is being separated from others boiling considerably lower (butanes) and higher (hexanes) but, if that were really the case, the curves for subsequent fractionations should become more horizontal (indicating greater purity) at about that temperature, and more vertical at, say, 25° and 40°.

Instead of this, they become less horizontal at about 33° and the next curve, IV, approximates to a straight line between about 30° and 37°. This change is a sure sign that there are at least two substances present boiling at temperatures not very far apart, but little or no light is thrown on the actual boiling points.

Curve VII is fairly straight from about 28·5° to about 35·5° but then becomes much more horizontal, terminating at 36·4°. This seems to indicate that the less volatile component boils at a temperature not far from 36° (the hexanes having now been almost completely eliminated).

. Curve X is distinctly more vertical in the middle; it becomes nearly horizontal above and terminates at 36·3°, showing that the higher boiling point is a little over 36·0°. As the curve is more horizontal below than No. VII there is evidence that the second component must boil not far from 28°.

The upper extremity of the curve XIII is perfectly horizontal at 36·3° and the true boiling point of the less volatile component (normal pentane) is thus established; the form of the lower part of the curve indicates that the boiling point of the more volatile component (isopentane) must be very close to 28° and further fractionation showed that it is really 27·95°.

Hexamethylene in American Petroleum.—When, in a complex mixture, one component is present in relatively very small quantity, it may very easily be overlooked, and it is only by keeping a careful record of the results of the fractionations and especially of the values $\Delta w/\Delta t$, or by plotting the weights of distillate against the temperature, that the presence of such substances can be detected.

As an example, consider the distillate from American petroleum coming over between 66° and 91° (6). In this case, again, no satisfactory result could be obtained without the use of a very efficient still-head. The results of the 5th, 6th, and 7th fractionations between the above-named limits of temperature and of the 12th fractionation between 74° and 85° are given in Table 41. The fifth fractionation gave high values of $\Delta w/\Delta t$ from 61·5° to 70°, with a maximum at 67° to 68°. Above 70° the ratios fell to a minimum at 78° to 84° and then rose again. Now, if there were no substance present between the hexanes and heptanes, the ratio $\Delta w/\Delta t$ would become smaller each time at the intermediate temperatures, and notably at about 80°; but it will be seen that in the next fractionation the ratio was slightly higher instead of lower. The number of fractions in the neighbourhood of 80° was therefore increased, and in the 7th fractionation a maximum value of $\Delta w/\Delta t$ was observed at about this temperature in addition to that due to normal hexane near 69°.

TABLE 41.

\multicolumn{4}{c}{V.}	\multicolumn{4}{c}{VI.}	\multicolumn{4}{c}{VII.}	\multicolumn{4}{c}{XII.}
F	t	Δw	$\dfrac{\Delta w}{\Delta t}$	F	t	Δw	$\dfrac{\Delta w}{\Delta t}$	F	t	Δw	$\dfrac{\Delta w}{\Delta t}$	F	t	Δw	$\dfrac{\Delta w}{\Delta t}$
1	(66″) 67	85·9	85·9	1	(66″) 67	74·9	74·9	1	(66·0) 66·9	49·8	55·3				
2	68	94·1	94·1	2	68	82·6	82·6	2	67·7	62·3	77·9				
								3	68·4	70·0	100·0				
4	69	81·1	81·1	4	69	103·0	103·0	4	69·0	68·8	114·7				
5	70	73·3	73·3	5	70	78·3	78·3	5	69·8	81·9	102·4				
6	72	64·6	32·3	6	71·5	52·2	34·8	6	71·0	46·9	39·1				
								7	72·5	45·4	30·3				
8	78	81·4	13·6	8	76·0	69·7	15·5	8	76·0	20·5	5·9		(74·0)		
												9	78·8	12·7	2·7
10	84	62·5	10·4	10	82·0	64·8	10·8	10	79·5	88·1	10·9	10	80·0	16·4	13·7
												11	80·5	24·3	48·6
												12	81·1	12·1	20·2
												13	83·0	11·5	6·1
14	89	66·0	13·2	14	87	58·9	11·8	14	85·0	72·9	13·3	14	85·0	7·1	3·5
15	91	39·9	20·0	15	90	35·3	11·8	15	88·0	23·4	7·8				

Further fractionation of the distillates coming over between 72° and 88° showed that there was really a substance present boiling not far from 80°, as will be seen from No. 12. It might be supposed that this substance was benzene (b.p. 80·2°) ; but that could not have been the case, for a 10 or 20 per cent. solution of benzene in hexane boils almost constantly at nearly the same temperature as hexane itself, and, when American petroleum is distilled, almost the whole of the benzene comes over below 70°, mostly from about 63° to 68°. Moreover, in this case the aromatic hydrocarbons had been removed by treatment with nitric and sulphuric acids.

The quantity of liquid was too small to allow of a fraction of quite constant boiling being obtained, though considerable improvement was effected.

Hexamethylene in Galician Petroleum.—Later on, a large quantity of Galician petroleum was fractionated by Miss E. C. Fortey (7), who obtained a considerable amount of liquid boiling quite constantly at 80·8°. A chemical examination of the liquid led to the conclusion that it consisted of pure hexamethylene (8) (9) ; but it was afterwards found (10) that it could be partially, but not completely, frozen in an ordinary freezing mixture, and it was necessary to resort to fractional crystallisation to separate the hexamethylene (b.p. 80·85°) in a pure state. It is evident that there is another hydrocarbon present in small quantity, no doubt a heptane, and that the two substances cannot be separated by fractional distillation.

Mixtures of Constant Boiling Point.—Reference has been made in Chapter IV. to the formation of mixtures of constant (minimum or maximum) boiling point. When a liquid contains two components which are capable of forming a mixture of constant boiling point, it is not possible to separate both components; all that can be done is to separate that component which is in excess from the mixture of constant boiling point, and even this is not possible when the boiling points are very near together. Thus, no amount of fractionation with the most perfect apparatus would make it possible to separate either pure normal hexane or pure benzene from a mixture containing, say, 2 per cent. of benzene (6), because the boiling points of hexane and of the binary mixture differ, in all probability, by less than 0·1°. On the other hand, if the original mixture contained, say, 50 per cent. of benzene, a small quantity of that component could be separated in a pure state from the mixture of constant boiling point (11).

American petroleum contains a relatively small amount of benzene, and the whole of it comes over with the hexanes; but Russian petroleum is much richer in aromatic hydrocarbons, and, consequently, some of the benzene comes over at its true boiling point, only a portion of it distilling with the hexanes.

When a liquid contains three components which are not closely related to each other it may happen that both a ternary and a binary mixture of constant boiling point are formed on distillation. In that case it is only possible to separate one of the original components in a pure state. These points will be considered more fully in Chapters XIII. and XV.

REFERENCES.

1. Kreis, "Comparative Investigations on the Methods of Fractional Distillation," *Liebig's Annalen*, 1884, **224**, 259.
2. Mendeléeff, A paper, without title, on the fractional distillation of Baku Petroleum, *Journ. Russ. Phys. Chem. Soc.*, Protok., 1883, 189.
3. Barrell, Thomas and Young, " On the Separation of Three Liquids by Fractional Distillation," *Phil. Mag.*, 1894, [V.], **37**, 8.
4. Young and Thomas, "Some Hydrocarbons from American Petroleum. I. Normal and Iso-pentane," *Trans. Chem. Soc.*, 1897, **71**, 440.

5. Young, "Experiments on Fractional Distillation," *Journ. Soc. Chem. Industry*, 1900, **19**, 1072.
6. Young, "Composition of American Petroleum," *Trans. Chem. Soc.*, 1898, **73**, 905.
7. Fortey, "Hexamethylene from American and Galician Petroleum," *Trans. Chem. Soc.*, 1898, **73**, 932.
8. Baeyer, "On the Hydro-derivatives of Benzene," *Berl. Berichte*, 1893, **26**, 229 ; "On the Reduction products of Benzene," *Liebig's Annalen*, 1893, **278**, 88.
9. Markownikoff, "On the presence of Hexanaphthene in Caucasian Naphtha," *Berl. Berichte*, 1895, **28**, 577 ; "On some new constituents of Caucasian Naphtha," *ibid.*, 1897, **30**, 974.
10. Young and Fortey, "The Vapour Pressures, Specific Volumes and Critical Constants of Hexamethylene," *Trans. Chem. Soc.*, 1899, **75**, 873.
11. Jackson and Young, "Specific Gravities and Boiling Points of Mixtures of Benzene and Normal Hexane," *Trans. Chem. Soc.*, 1898, **73**, 923.

CHAPTER VIII

THEORETICAL RELATIONS BETWEEN THE WEIGHT AND COMPOSITION OF DISTILLATE

Application of Brown's Formula. (1)—It has been pointed out in previous chapters

1. That the vapour pressures of mixtures of two closely related compounds—and rarely of others—are represented with small error by the formula $P = \dfrac{mP_A + (100 - m)P_B}{100}$;

2. That when this formula holds good, the composition of the vapour from any mixture is given, approximately at any rate, by Brown's formula $\dfrac{x_B}{x_A} = c\,\dfrac{W_B}{W_A}$, where the constant c does not differ greatly from the mean ratio of the vapour pressures of the pure substances at temperatures between their respective boiling points. We may conclude,

then, that Brown's formula can be used without much error for mixtures of two closely related substances, and it is probable that the two formulæ referred to above, when suitably modified, are applicable also to mixtures of three or more closely related substances.

Mixtures of Two Components.—Taking first the case of mixtures of two liquids, Brown's formula may be written

$$\frac{d\xi}{d\eta} = c\,\frac{\xi}{\eta}$$

where ξ = residue of liquid B at any instant, and η = residue of liquid A at the same instant.

Taking L and M as the weights of B and A originally present and $L + M = 1$, we obtain by integration

$$(M^c/L)y\{c + (1-c)y\}^{c-1} = c^c(1-x)^{c-1}(1-y)^c$$

where y = quantity of the more volatile liquid B in unit weight of the distillate coming over at the instant when x is the quantity of liquid distilled. By means of this equation the changes of composition that take place in the course of a distillation may be traced, and the variation in the composition of the distillate represented graphically.

To take a very simple case, suppose that $c = 2$ and that $L = M = \frac{1}{2}$.

In the diagram (Fig. 30), the amounts of distillate that have been collected are represented as abscissæ, and the relative quantities of the two liquids, A and B, in the distillate as ordinates.

It will be seen that the composition of the distillate alters slowly at first, then more and more rapidly, also that while the first portion of the

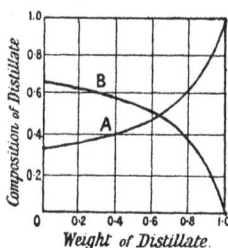

FIG. 30.—$B = 0.5$, $A = 0.5$, in original mixture.

distillate contains a considerable amount of the less volatile substance, A, the last portion is very nearly free from the lower boiling component, B. These points are fully confirmed by experiment, and an explanation is afforded of the fact that it is much easier to separate the less volatile than the other component in a pure state.

By fractionating a few times in the ordinary way, collecting the distillates in six or eight fractions, we shall have a large excess of B in the first fraction and a still larger excess of A in the last.

Suppose, now, that two of these fractions, one containing B and A in the ratio of $9 : 1$ and the other in the ratio $1 : 9$, are distilled separately and completely; the results will then be represented by Figs. 31 and 32. Again, it will be seen

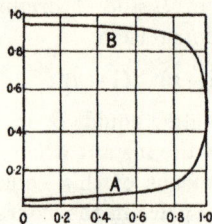

Fig. 31.—$B = 0.9$, $A = 0.1$, in mixture distilled.

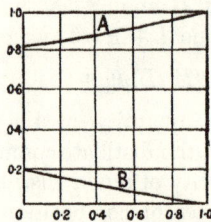

Fig. 32.—$B = 0.1$, $A = 0.9$, in mixture distilled.

that the first tenth of the distillate from the first of these fractions is not so rich in B as the last tenth from the second fraction is in A; the purification of A still proceeds more rapidly than that of B.

In making use of this formula it is assumed that no condensation (and therefore no fractionation) goes on in the still-head, but that the vapour reaches the condenser in the same state as when first evolved from the liquid in the still. By using an improved still-head (Chapters X to XII) a more rapid separation would be effected.

Mixtures of Three Components.—If we have a mixture of three closely related substances, C, B and A, it may be conjectured that the proportion of the three substances in the vapour at any instant is the same as that of the weights of the three substances in the residue in the still, each weight being multiplied by a suitable constant, which is approximately proportional to the vapour pressure of the corresponding liquid.   Here again formulæ may be obtained

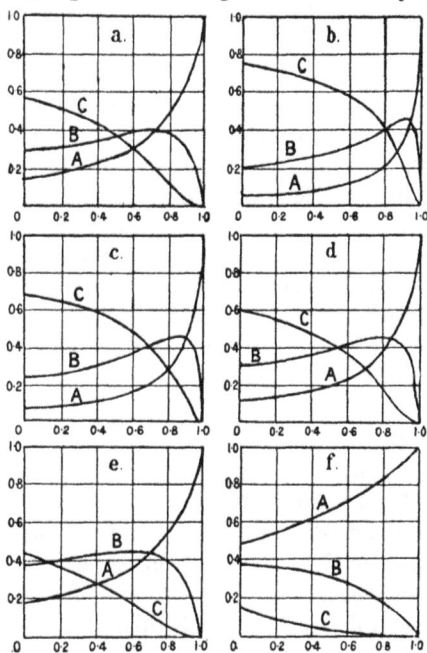

Fig. 33 (a to f).—Distillation of mixtures containing three components, A, B, and C.

by integration, which enable us to follow the course of a distillation.

Taking the three constants as $c = 4$, $b = 2$, $a = 1$—which are roughly proportional to the vapour pressures of methyl, ethyl and propyl acetates at the same temperature—and the original weights of the three liquids C, B and A as $L = M = N = \frac{1}{3}$, Fig 33 a represents the first distillation.

If the distillate were collected in five equal fractions they would have the following composition :—

TABLE 42.

	C or L.	B or M.	A or N.
IIa	0·543	0·300	0·157
IIβ	0·47	0·33	0·20
IIγ	0·37	0·365	0·265
IIδ	0·22	0·39	0·39
IIϵ	0·047	0·265	0·687

It will be seen that while the last fraction is more than twice as rich in A as the original mixture and is nearly free from C, and the amount of C in the first fraction is 0·543, as against 0·333 in the original mixture, the fraction richest in B, the fourth, contains only 0·39 of that substance. It is evident, therefore, that the middle substance is the one that is the most difficult, and the least volatile substance the one that is easiest to separate. If these five fractions were separately distilled we should get the results indicated in the curves, Fig. 33, b, c, d, e, f.

It is to be noted that in IIa, which is richest in the most volatile component, C, the amount of C rises from 0·543 to about 0·72 in the first fifth of the distillate. In IIϵ, which is richest in the least volatile liquid A, the amount of A rises from 0·687 to about 0·9 in the last fifth of the distillate. On the other hand, in IIδ, which is richest in B, the middle liquid, the improvement is merely from 0·39 to rather less than 0·45 in the third fifth of the distillate.

If the fractions richest in C, B and A respectively were redistilled we should have the following result for the best distillates :—

Improvement in C from . . 0·7 to 0·82
,, B ,, . . 0·45 ,, 0·50
,, A ,, . . 0·9 ,, 0·98

Thus, the improvement in B continues to be very slow, and even when a fraction very rich in B is redistilled the improvement is very slight. Thus, on redistilling a mixture containing 0·02 of C, 0·96 of B and 0·02 of A, the amount of B advances only from 0·96 to 0·97 in the most favourable part of the distillate; it is thus clear that there must be far greater difficulty in separating the middle substance than either of the others from a mixture.

It is, however, not difficult to obtain fractions rich in C but free from A, and also rich in A but free from C, and the best method of obtaining B in a pure state is to carry out a series of preliminary fractionations in the manner described for methyl, ethyl and propyl acetates (p. 128) until all the fractions coming over below the boiling point of B contain only C and B, and all those coming over above that temperature contain only B and A. We have, then, only to deal with the separation of two components from the mixtures, and these separations will already be far advanced.

REFERENCE.

1. Barrell, Thomas and Young, "On the Separation of Three Liquids by Fractional Distillation," *Phil. Mag.*, 1894, [V], **37**, 8.

CHAPTER IX

Data required.—If, for mixtures of any two substances, the curve representing the relations between boiling point and molecular composition has been constructed, and if the relation between the composition of the liquid mixture and of its vapour is known, the boiling point of the distillate may be read from the curve. We have seen that if the two substances are closely related we may safely assume that the boiling points of mixtures may be calculated from the vapour pressures of the components, and that Brown's formula, taking the mean ratio of the vapour pressures for the value of the constant, c, gives the relation between the composition of liquid and vapour with fair accuracy.

Benzene and Toluene.—In the case of benzene and toluene, the boiling points of mixtures have been found to agree very closely with those read from the theoretical curve, but the relation between the composition of liquid and vapour has not been experimentally determined. The mean ratio of the vapour pressures at temperatures between 80° and 110° is about 2·5, and we may probably assume that the relation is expressed with sufficient accuracy by the formula $x_B/x_A = 2\cdot5\ W_B/W_A$.

In Table 43 are given—

1. The molecular percentages of benzene in mixtures which boil at the temperatures in the first column,

2. The corresponding molecular percentages of benzene in the distillate, calculated by means of Brown's formula,

3. The boiling points of these distillates read from the curve,

4. The differences (Δt) between the boiling points of liquid and distillate.

TABLE 43.

Benzene and Toluene.

Boiling point of liquid	Molecular percentage of Benzene.		Boiling point of dis- tillate.	Δt	Boiling point of liquid.	Molecular percentage of Benzene.		Boiling point of dis- tillate.	Δt
	Liquid.	Dis- tillate.				Liquid.	Dis- tillate.		
81°	96·3	98·5	80·55°	0·45°	96°	38·3	60·8	89·35°	6·65
82	91·5	96·4	81·0	1·0	98	32·2	54·3	91·15	6·85
84	82·4	92·1	81·95	2·05	100	26·4	47·3	93·2	6·8
86	73·8	87·6	82·85	3·15	102	21·0	39·9	95·5	6·5
88	66·0	82·9	83·9	4·1	104	15·8	31·9	98·1	5·9
90	58·4	77·8	85·05	4·95	106	10·8	23·2	101·2	4·8
92	51·3	72·5	86·3	5·7	108	6·0	13·8	104·75	3·25
94	44·6	66·8	87·8	6·2	110	1·4	3·4	109·1	0·9

Isopentane and n-*Pentane.*

	Isopentane.					Isopentane.			
28·5°	92·5	94·25	28·38°	0·12°	33°	36·35	43·15	32·40°	0·60
29	85·95	89·05	28·76	0·24	34	25·0	30·70	33·49	0·51
30	72·9	78·15	29·59	0·41	35	13·95	17·75	34·65	0·35
31	60·3	66·90	30·48	0·52	36	3·15	4·15	34·91	0·08
32	48·0	55·10	31·41	0·59					

Isopentane and Normal Pentane.—The corresponding data for mixtures of isopentane and normal pentane are given in the second part of Table 43, but in this case neither the boiling points of mixtures nor the relation between the composition of liquid and of vapour have been directly

determined. The theoretical boiling point curve differs
very slightly from a straight line, the maximum deviation,
D_1, being $0.28°$ when the molecular percentage of normal
pentane, m_D, is 51.5 (p. 59). The ratio of the vapour
pressures at $30°$ is 1.33, and this has been taken as the
constant in Brown's formula.

Application to Distillation of Benzene and Toluene.
—Let us consider first the case of benzene and toluene. In
the first four fractionations, details of which are given on
p. 118, the range of temperature for most of the fractions was
$3°$. For fraction 6, collected from $92.2°$ to $95.4°$, the middle
temperature is $93.8°$, and from Table 43 we see that the distillate from a mixture boiling at $93.8°$ would have a boiling
point about $6.2°$ lower, or $87.6°$. That is to say, if this fraction
were distilled in such a manner that no condensation could
take place in the still-head, it would begin to boil at $87.6°$, but,
as a matter of fact, there was some condensation which
would lower the temperature to some extent at first. On
the other hand, in the actual fractionation, the fractions,
with the exception of the first, were not distilled separately
but were added to the residues left in the still. Thus the
fraction that came over from $92.2°$ to $95.4°$ was added
to the residue from No. 5 which was boiling at $92.2°$, and
was therefore richer in toluene, and the boiling point
would therefore be somewhat higher than if the fraction
were distilled alone. We may perhaps suppose that the
two disturbing factors would about counterbalance each
other, and that the mixture would actually begin to boil
at about $87.5°$. The temperature ranges below $92.2°$ to
$95.4°$ were as follows: No. 5, $89.2°$ to $92.2°$; No. 4, $86.2°$
to $89.2°$, and it is clear that a considerable amount of
distillate would be collected below $89.2°$ in receiver No. 4.

In the sixth fractionation (p. 124) the corresponding
temperature ranges were—No. 4, $84.0°$ to $87.0°$; No. 5, $87.0°$
to $90.5°$; No. 6, $90.5°$, to $95.4°$. The middle temperature for

No. 6 would be 92·95°, and the distillate from this would begin to boil at about 87·0° or 5·95° lower. In this case none of the distillate would be collected in No. 4, and receiver No. 5 might be (and actually was) left in position when No. 6 fraction was added to the residue in the still and the mixture was redistilled.

In the thirteenth fractionation the range of No. 6 had been increased to 14°, from 81·4° to 95·4°. The middle temperature would be 88·4°, and the boiling point of the distillate would be, roughly, 88·4° − 4·2° = 84·2°, so that no distillate would be collected in No. 5, and there would be no object in continuing the fractionation in the same way as before. This was, in fact, found to be the case, the temperature rising at once above 81°. The fractions above and below 95·4° were therefore treated separately as described on p. 128.

Application to Distillation of Pentanes.—Let us now consider the behaviour of a mixture of isopentane and normal pentane. The difference between the boiling points is 8·35° and the middle temperature is 32·6°. If the distillate were collected in the same number of fractions as in the case of benzene and toluene, the range of the middle ones would be about 0·84°, say, 0·8°. We should then have for the corresponding fractions—No. 5, 31·0°—31·8° ; No. 6, 31·8°—32·6°.

The middle temperature for No. 6 would be 32·2°, and the boiling point of the distillate from this would be 32·2°—0·6° = 31·6°. Very little would therefore be collected even in No. 5. The temperature range would have to be reduced to 0·4° [No. 5, 31·8°—32·2° ; No. 6, 32·2°—32·6°] in order that the distillate from No. 6 might begin to boil at the initial temperature of No. 5, and that would mean that the number of fractions would have to be doubled to begin with, and that it would not be possible to increase the temperature range of the middle fractions beyond a very small amount.

It will thus be seen that, even without considering the excessive loss by evaporation of such volatile liquids, the separation of isopentane from normal pentane with an ordinary still-head would be practically impossible. It is only by using a greatly improved still-head that such a difficult separation is rendered possible.

CHAPTER X

MODIFICATIONS OF THE STILL-HEAD

Object of Modifications.—It is evident from what has been stated in Chapter VII that the process of fractional distillation with an ordinary still-head is frequently an exceedingly tedious one, and many attempts have been made to modify the still-head in such a manner as to bring about a more complete separation in a single operation.

The idea of improving the still-head is by no means a new one, for in a book entitled *Philosophorum, seu liber de Secretis Naturæ*, by Philip Ulstadius, 1553, there is an illustration of a five-headed still (Fig. 34). When dilute spirit was distilled through this "alembic" it is clear that the alcohol collected in the uppermost receiver must have been considerably stronger than that in the lowest.

FIG. 34.—Five-headed still described by Ulstadius, 1553.

Great progress was made in the improvement of the still-head in commerce—notably for the better separation of alcohol from weak spirit—before any advance was made in the laboratory and, indeed, the improved still-heads first employed in the laboratory were, for the most part, merely adaptations of those already in use on the large scale.

It may, however, be convenient to consider first those forms of apparatus which are suitable for laboratory purposes.

The object of the modifications which have been made in the still-head is to bring about a more complete separation of the components of the mixture in a single distillation, a process of fractionation—more or less effective according to the form of apparatus and the rate of distillation—taking place in the still-head (1).

Inefficiency of Plain Vertical Still-Head.—Of all possible forms, the plain vertical still-head is the least efficient. During the distillation, as the vapour rises up the cylindrical tube, the outer parts of it come in contact with the condensed liquid flowing down the sides of the tube. Since the tube is constantly losing heat by radiation and by conduction to the surrounding air, this liquid is slightly cooled and fresh condensation constantly goes on, the outermost layer of vapour probably condensing almost completely without much change of composition. The central portion of vapour rises rapidly up the tube and can only reach the liquid by diffusion or by convection currents, and much of it may pass through the still-head without reaching the liquid at all ; the condensed liquid, on the other hand, flows rapidly down the vertical walls of the tube, back to the still. In order, however, that a satisfactory separation of the components of the mixture may take place, as much of the vapour as possible should be brought into intimate contact with the condensed liquid, so that a state of equilibrium, as regards composition, may be brought about.

Brown's Formula.—Let us suppose, to facilitate the consideration of what occurs, that we are dealing with a mixture of two closely related compounds, to which Brown's formula $\dfrac{x_B}{x_A} = c\dfrac{W_B}{W_A}$ is applicable, and that x and W refer to the weights of vapour and liquid respectively. Also let us suppose that, in the first place, equal weights of the two substances are taken, so that $W_1 = W_2$.

Then the composition of the first small quantity of vapour formed will be given by the equation $\dfrac{x_B}{x_A} = c$ and the percentage of B in the vapour will be $\dfrac{c \times 100}{c + 1}$.

If a distillation bulb with exceedingly short still-head be employed, this will practically be the percentage of B in the first portion of distillate.

Theoretically Perfect Still-Head.

—Suppose, now, that in the lower part of a long and theoretically perfect still-head condensation goes on, and that the condensed liquid remains in this part of the tube until its weight is, say, 100 times as great as that of the residual uncondensed vapour in the same part of the tube, the total weight of liquid and vapour being, say, 1/1000 of that of the mixture taken. We may then assume

1. That the composition of the liquid and vapour (taken together) in the still-head will be practically the same as that of the vapour first formed, and

2. That the composition of the condensed liquid will not differ sensibly from this, or $\dfrac{W'_B}{W'_A} = c$.

If the residual vapour is brought into thorough contact with the condensed liquid, its composition will practically be given by the equation $\dfrac{x'_B}{x'_A} = c \times c = c^2$, and the percentage of B in it will be $\dfrac{c^2 \times 100}{c^2 + 1}$.

If we suppose, as a theoretical case, that the still-head is divided into n sections, and that the distillation goes on in such a manner that perfect equilibrium between vapour and condensed liquid is established in each section, and, lastly, that the condensed liquid in any section has the same composition as the vapour in the section below, then we should have, for the composition of the vapour in the nth section, when the distillate begins to come over, $\dfrac{x_B}{x_A} = c^{\,n+1}$.

If we take $c = 2\cdot5$, as in the case of benzene and toluene, and start with equal weights of the components, the composition of the vapour in the still and in the first nine sections would be as follows :—

TABLE 44.

No. of Section.	Percentage of B.	No. of Section.	Percentage of B.
Still	71·43	5	99·59
1	86·21	6	99·84
2	94·55	7	99·93
3	97·51	8	99·97
4	98·99	9	99·99

Such an arrangement is not realisable in practice, but the attempt should be made to approach as closely to it as possible.

Characteristics of a good Still-Head.—In choosing or constructing a still-head, the first point to be considered is its efficiency in separating the components of a mixture. It frequently happens, however, that the quantity of liquid available is small and, in any case, when a series of fractional distillations has to be carried out, some of the fractions eventually become very small : in such a case, of two still-heads of equal efficiency, that one is the more useful which allows of the distillation of the smaller quantity of liquid. Now there is necessarily, at any moment, a certain amount of condensed liquid in every efficient still-head and, obviously, the smallest quantity of a substance that can be usefully distilled must be considerably greater than that of the liquid and vapour in the still-head. It is therefore of great importance to construct the still-head in such a manner that—consistently with efficiency—the quantity of condensed liquid in it at any moment shall be as small as possible. It is also important that, after the removal of the source of heat, this liquid shall return as completely as possible to the still.

Among other points to be considered are (a) ease of construction, (b) freedom from liability to fracture, (c) convenience in handling.

Comparison of Still-Heads. Mixture Distilled.—In comparing the efficiency and usefulness of different still-heads it is necessary always to distil a mixture of the same composition. Equal weights of pure benzene and toluene may conveniently be employed.

Rate of Distillation.—It is necessary also that the rate of distillation should always be the same, and it is best to collect the distillate at the rate of one drop per second. A good plan is to have a seconds pendulum—a weight attached to a string 39·1 inches or, say, 1 metre long, serves the purpose very well—swinging behind the receiver, so that the drops may be easily timed.

That the rate of distillation does greatly influence the separation was pointed out by Brown (2), and is well shown by Table 45, in which are recorded the results of three distillations, each of 50 grams of the benzene-toluene mixture, at the rate of 30, 60, and 120 drops per minute. An improved still-head was used, and the distillate was collected in eleven fractions, the twelfth consisting of the residue (calculated by difference) after the temperature had reached the boiling point of pure toluene. For convenience, the results are stated as percentages.

TABLE 45.

Temperature range.	Number of drops of distillate per minute.		
	30	60	120
	Percentage weight of distillate.		
80·2— 83·2°	0·8	0·6	0·6
83·2— 86·2	29·6	21·8	10·8
86·2— 89·2	9·8	14·3	20·2
89·2— 92·3	5·6	7·8	12·6
92·3— 95·4	3·8	5·0	7·0
95·4— 98·5	3·0	4·7	5·4
98·5—101·6	2·8	4·0	5·0
101·6—104·6	3·0	3·6	5·0
104·6—107·6	4·3	5·2	5·6
107·6—110·0	6·2	7·3	9·2
110·0—110·6	11·6	9·8	11·1
Pure toluene by difference .	19·5	15·9	7·6*
	100·0	100·0	100·0

* Temperature barely reached 110·6° ; residue not quite pure toluene.

It will be seen that the separation is greatly improved by diminishing the rate of distillation.

Explanation of Tables.—In Tables 46 to 50, which show the relative efficiency and usefulness of different still-heads, the following data are given.

1. The vertical height from the bottom of the still-head to the side delivery tube.

2. The final temperature for each fraction.

3. The percentage weight of distillate.

4. The weight of vapour and liquid in the still-head during the distillation when pure or nearly pure toluene was coming over. This was arrived at by continuing the distillation as nearly as possible to the last drop and weighing the liquid, when cold, in the flask (still) and—if necessary and possible —in the still-head. From this amount was subtracted the constant weight of liquid and vapour in the flask itself, which was estimated to be 0·85 gram.

For the very long still-heads, which could not conveniently be weighed, the estimated loss by evaporation $+0.85$ gram was subtracted from the difference between the original weight of the mixture and the sum of the weights of the fractions. The actual loss by evaporation, when measured, varied from 0·00 to 0·65 gram, and was greater for the large than the small still-heads.

The Plain Vertical Still-Head.—Although the plain vertical still-head is less efficient than any other, yet a certain amount of fractionation does take place in it, and it may be well to consider the effect of altering (a) its length, (b) its internal diameter. The influence of such alterations is shown in Table 46.

TABLE 46.

PLAIN VERTICAL STILL-HEADS.

Vertical height in cm.	12	62	120	From 62 to 70 cm.				
Internal diameter in mm.	From 14 to 15·5 mm.			5·1	8·0	14·0	19·4	25·7
Final temperature.	Percentage weight of distillate.							
83·2°	...	...	...	...	...	...	...	...
86·2	...	0·1	1·0	0·5	0·4	0·1	1·3	0·2
89·2	2·1	12·8	19·4	22·2	16·2	12·8	15·2	20·0
92·3	28·4	21·5	16·6	16·5	18·3	21·5	19·4	18·8
95·4	18·0	11·0	10·6	9·2	11·7	11·0	11·2	9·2
98·5	10·9	8·9	7·2	7·2	8·7	8·9	8·6	7·2
101·6	8·6	7·4	6·5	5·8	7·5	7·4	6·1	4·7
104·6	7·2	6·2	6·0	5·5	5·8	6·2	5·6	5·9
107·6	7·2	7·3	6·1	5·9	6·9	7·3	6·6	6·4
110·0	6·8	8·4	7·8	7·1	8·2	8·4	9·5	10·0
110·6	5·9	9·0	8·1	8·9	9·5	9·0	8·0	7·2
Pure toluene by difference .	4·9	7·4	10·7	11·2	6·8	7·4	8·5	10·4
	100·0	100·0	100·0	100·0	100·0	100·0	100·0	100·0
Weight of liquid and vapour in still-head	0·3	1·55	3·55	1·05	1·15	1·55	2·35	3·15

Influence of Length.—In the first three distillations the diameter of the still-head was nearly the same, but the third tube was ten times as long as the first. As might be expected, the efficiency is improved by this alteration, but the weight of liquid and vapour also increases and is, roughly, proportional to the length.

Influence of Width.—For tubes of approximately equal length the efficiency is smallest when the diameter is rather less than 14 mm., and rises when it is either increased or diminished, as will be seen from the results of the last five distillations. On the other hand, the weight of liquid and vapour in the still-head increases with the diameter of the tube, and it is therefore clearly more advantageous, in making a plain still-head, to use very narrow rather than very wide tubing. It will be noticed also that the narrowest tube gives a better result than the tube of medium diameter nearly twice as long, though the weight of liquid and vapour is only 1·05 as compared with 3·55 grams.

The diameter of the tubing cannot, however, be diminished beyond a certain amount, depending on the nature of the liquid distilled, especially on its boiling point, for if the tube is too narrow, the condensed liquid will unite into columns which will be driven bodily upwards. This blocking is much less liable to occur if a short piece of wider tube is sealed to the bottom of the narrow one. There must also be a wider piece at the top to admit the thermometer.

Condensation in the still-head may be diminished by covering it with cotton wool or any other non-conducting material, but while the amount of liquid in the tube is thus diminished, so also is the efficiency. The device is, however, occasionally useful, for if the liquid in the tube is just on the point of joining up into columns, the blocking may be prevented by slightly diminishing the amount of condensation.

Modifications of the Still-Head.—It has been pointed out that the inefficiency of the plain vertical still-head is due to the want of thorough contact between the vapour and the condensed liquid, owing, firstly, to the central portion of the vapour passing rapidly up the tube and possibly never meeting with the liquid at all, and, secondly, to the con-

densed liquid flowing very rapidly back to the still.
Consequently, any modification that brings about better
admixture of the ascending vapour or that retards the
down-flow of condensed liquid should therefore increase its
efficiency.

Sloping Still-Head.—The simplest and most obvious alteration that can be made is to change the slope of the tube
so as to retard the down-flow of liquid. This may be done
by bending the tube near the top and bottom so that its ends
remain vertical, while the middle part slopes very gently
upwards.

It will be seen from Table 47 that, by altering a tube in

TABLE 47.

Nature of Still-head.	Vertical.	Sloping.	Spiral.	Plain vertical.	With rod and 20 discs.	Same as last with constrictions.
Vertical height in cm.	70	28·5	32	62 to 63		
Internal diameter.	8 mm.			14 mm.		
Final temperature.	Percentage weight of distillate.					
83·2°	...	...	...	...	1·4	2·0
86·2	0·4	3·5	8·0	0·1	18·1	20·5
89·2	16·2	24·8	22·6	12·8	15·8	14·4
92·3	18·3	14·0	11·6	21·5	9·7	8·6
95·4	11·7	9·3	8·8	11·0	6·1	5·8
98·5	8·7	6·7	5·5	8·9	4·2	3·6
101·6	7·5	5·6	5·0	7·4	3·8	3·4
104·6	5·8	5·4	5·5	6·2	4·2	4·2
107·6	6·9	6·3	6·0	7·3	4·3	4·2
110·0	8·2	8·2	7·4	8·4	8·0	7·4
110·6	9·5	9·0	10·7	9·0	6·6	7·5
Pure toluene by difference	6·8	7·2	8·9	7·4	17·8	18·4
	100·0	100·0	100·0	100·0	1·000	100·0
Weight of liquid and vapour in still-head .	1·15	1·10	0·95	1·55	2·35	2·2

the manner described, a notable improvement is effected,
especially as regards the separation of the benzene, and it is
remarkable that, although the down-flow of liquid is retarded,
the quantity of liquid in the still-head is slightly diminished.

Spiral Still-Head.—The efficiency is further improved by
bending the sloping portion of the tube into the form of a
spiral, probably because a better admixture of the vapour is
thus produced ; by this device the amount of liquid in the
still-head is still further reduced.

"Rod and Disc" Still-Head.—In the still-head shown
in Fig. 35a the down-flow of part of the condensed liquid is
greatly retarded by the discs on
the central glass rod, and this
liquid is protected from the
cooling action of the air ; at the
same time eddies and cross currents are produced in the ascending vapour. The increase
in efficiency with this apparatus
is very marked, as will be seen
from the second part of Table
47 ; it is easily constructed and
is very convenient to handle.
The quantity of liquid and vapour in the still-head is the same
as that in the plain tube of 19·4
mm. diameter, but the efficiency
is very much greater. On re-
moving the source of heat, the

FIG. 35.—The "rod and disc" still-
heads ; (a) without, (b) with
constrictions in outer tube.

liquid returns almost completely to the still, but when it is
of special importance to avoid all loss, it is advisable with
this and many other forms of still-head to disconnect the
condenser and to tilt the still and still-head, while hot, from
side to side so as to facilitate the back-flow of liquid. If that
is done, the whole of the liquid usually returns to the still.

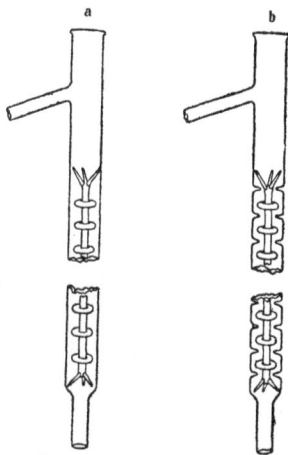

A slight further improvement, both in efficiency and as regards the quantity of liquid in the still-head, is effected by constricting the outer tube between the discs (Fig 35b). Better contact between the vapour and the condensed liquid on the outer tube is thus ensured.

Bulb Still-Heads.—A vertical tube with a series of bulbs blown on it was recommended by Wurtz (3). For a given diameter of tube and bulbs, the greater the number of bulbs the higher is the efficiency, but the greater also is the quantity of liquid and vapour in the still-head (Table 48).*

TABLE 48.

Nature of still-head.	3 bulbs.	7 bulbs.	13 bulbs.	"Pear" still-head, 13 bulbs.
Vertical height in cm.	26	42	66	62
Final temperature.	Percentage weight of distillate.			
83·2°	...	0·2	1·4	3·0
86·2	2·2	14·2	24·4	26·2
89·2	18·8	18·0	10·2	11·0
92·3	18·4	12·0	7·9	5·8
95·4	9·4	5·1	6·0	5·2
98·5	7·8	6·4	3·7	2·4
101·6	6·7	4·6	3·6	2·4
104·6	5·8	4·8	3·4	2·8
107·6	6·4	5·6	3·4	3·6
110·0	8·0	6·2	6·2	5·0
110·6	10·0	10·4	8·0	11·0
Pure toluene by difference	6·5	12·5	21·8	21·6
	100·0	100·0	100·0	100·0
Weight of liquid and vapour in still-head	0·95	1·75	3·55	2·6

* Neither the experimental results obtained by Kreis (4) nor the conclusions he deduces from them can be accepted, for he obtained a very much worse separation of benzene with four bulbs than with two, though the separation of toluene was better. It was, however, according to his figures, not so good as with an ordinary distillation bulb.

The tube with thirteen bulbs was somewhat more efficient that the slightly shorter tube with rod and discs, but the quantity of liquid and vapour in the still-head was half as large again.

The "Pear" Still-Head.—The "Wurtz" still-head may be improved, to some extent in efficiency, but chiefly as regards the quantity of condensed liquid, by blowing pear-shaped instead of spherical bulbs on the tube (Fig. 36).

As a result of this alteration, the condensed liquid in any bulb, after flowing past the constriction, instead of spreading itself over the inner surface of the bulb below, mixing with the liquid condensed in that bulb and flowing down the sides with increasing velocity, collects on the depression in the bulb below and falls in drops near the middle of the bulb. The liquid on the inner surface of each bulb is merely the small

FIG. 36.—The "pear" still-head.

amount condensed in that bulb, and its velocity of down-flow is no greater in the bottom bulb than in the top one. The liquid, on the other hand, that collects on a depression (that is to say, the total quantity condensed in the part of the still-head above it) is brought well in contact with the ascending vapour in a part of the bulb that is less exposed to the cooling action of the air than any other. It is probable also that the eddies in the vapour are greater than in the ordinary bulb tube.

The "pear" still-head is more efficient than the "rod and disc" tube of the same length and possesses the same advantages except that it is somewhat more difficult to construct. It may be especially recommended for liquids of high boiling point.

The "Evaporator" Still-Heads.—Greater efficiency, for a given vertical height, and less condensation, for a given efficiency, is attained by the "evaporator" still-heads.

Original Form of "Evaporator" Still-Head.—The general form of the apparatus, as originally designed, is shown in Fig. 37. Each section consists of three separate parts—

1. The outer tube, A, of 22 to 24 mm. internal diameter, connected above and below with other sections, the length of each being about 10 or 10·5 cm. ;

2. An inner thin-walled tube, B, of 7·5 to 8 mm. internal diameter and 60 mm. long, open at each end and widened below into the form of a funnel which rests on the constricted part of A and may be prevented from fitting it too accurately by fusing three or four minute beads of glass to the rim of the funnel. Two large holes, (B'), are blown on the sides of B near the top ;

3. An intermediate tube, C, of about 14 mm. internal diameter and 50 mm. long, like a small inverted test tube. Above the tube C and attached to it by three glass legs, shown in Fig. 38, is a small funnel, C_1, which must be a little wider than the depression in the tube A just above it. The tubes B and C are centred and kept in position by the little glass projections shown in the same figure.

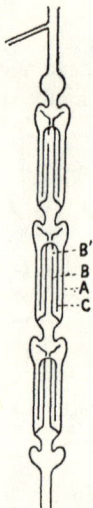

Fig. 37.
The "evaporator" still-head ; original form.

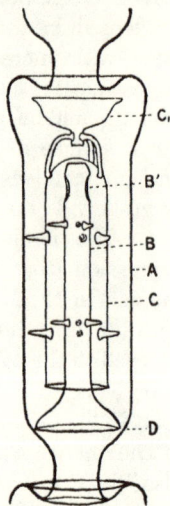

Fig. 38.—One section of the original " evaporator " still-head.

When the vapour first reaches a section from below, a large amount of condensation takes place, and the narrow passage, D, where the inner tube rests on the constriction in the outer one, becomes at once blocked by the condensed liquid. The vapour therefore rises up the inner tube, then passes down

between the inner and middle tubes and finally up again between the middle and outer tubes and so into the section
above.

The condensed liquid in any section collects together and
falls in drops from the depression in the section below into
the funnel; from this it falls on to the top of the middle
tube and spreads itself over its surface, falling again in drops
from the bottom of this tube and finally flowing through the
passage D.

Owing to gradual removal of the less volatile component,
the condensing point of a mixed vapour becomes lower and
lower as the vapour rises through the still-head; thus,
in any section, the vapour that rises through the inner
tube will be hotter than that which reaches the section
above. The condensing point of the vapour in the inner
tube must, indeed, be higher than the boiling point of the
liquid that falls from the little funnel, and when the two
components differ considerably in volatility and neither of
them is in great excess, evaporation of the liquid on the
middle tube may be easily observed. Under such conditions
there is a tendency for the drops of liquid from the funnel
to assume the spheroidal state, and the progress of a distillation is, indeed, rendered evident by the appearance or
disappearance of such drops.

With a pure liquid, the spheroidal drops are never seen
unless the quantity of liquid in the still is very small, when,
owing to superheating of the vapour, they may be formed
in the lowest section.

Modified " Evaporator " Still-Head.—A modification of
the "evaporator" still-head is shown in Fig. 39. In this,
the rather fragile funnel on the three legs is done away with,
and the top of the middle tube c is blown into a flattened
bulb c', on which the drops of liquid from the depression
above fall and collect into a shallow pool which soon over-
flows, and the liquid then spreads itself as before over the

surface of the tube. Spheroidal drops are not nearly so readily formed in this apparatus, but when rapid separation is taking place, the liquid flowing down the sides of the inverted tube breaks up into separate streams.

It will be seen from Table 49 that very good results are obtained with the evaporator still-heads of either form. The sections in the modified apparatus are shorter than in the original one, and for a given height of still-head the efficiency is greater, while, for a given efficiency, the weight of liquid and vapour in the still-head is usually slightly less.

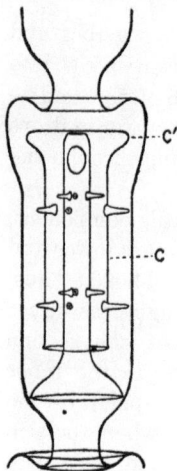

FIG. 39.—One section of the modified "evaporator" still-head.

When there are many sections, the inverted tube should be made somewhat shorter in the lower sections, so as to increase the vertical distance from the bottom of that tube to the bottom of the inner tube; in the modified still-head the vertical distance from the depression to the flattened top of the inverted tube should be somewhat increased in the lower sections.

The "Hempel" Still-head.—The great advantages of the "Hempel" apparatus (5) are simplicity and efficiency; on the other hand the amount of liquid in the still-head is excessive, and it is therefore unsuitable for the distillation of small quantities of liquid.

The still-head consists simply of a wide vertical tube, filled with glass beads of special construction, and constricted below to prevent the beads from falling out. A short, narrower, vertical tube with side delivery tube is fitted by means of an ordinary cork into the wide tube.

From the following table it will be seen that in efficiency

TABLE 49.

Nature of still-head.	"Evaporator" still-heads.						"Hempel," 200 large beads.
	Original.		Modified.				
Number of sections.	3	5	3	5	8	13	
Vertical height in cm.	57	77	46	62	78	131	58

Final temperature.	Percentage weight of distillate.						
81·2°	}15·4	6·5 }12·8		3·5	12·0	42·5*	}20·6
83·2		24·1		22·5	23·85	2·6	
86·2	19·6	10·0	21·15	12·35	6·5	1·7	15·4
89·2	6·2	3·8	7·5	5·6	2·9	1·1	7·2
92·3	5·1	2·3	4·8	3·25	2·15	0·65	3·4
95·4	3·5	1·6	3·35	1·95	1·4	0·5	3·2
98·5	2·9	1·7	2·7	1·3	1·15	0·55	2·2
101·6	2·6	1·5	2·3	1·8	1·05	0·5	1·8
104·6	2·4	1·8	2·5	1·5	1·15	0·45	2·9
107·6	3·8	2·9	3·5	2·45	1·6	0·9	3·2
110·0	5·8	4·8	6·5	4·2	3·95	1·95	6·8
110·6	10·8	8·2	10·5	11·4	9·6	2·85	8·4
Pure toluene by difference	21·9	30·8	22·4	27·9	32·7	43·75	24·9
	100·0	100·0	100·00	100·00	100·0	100·00	100·0
Weight of liquid and vapour in still-head	3·45	5·0	2·7	4·55	6·25	16·25	7·85

the Hempel tube, which contained 200 of the large beads now used, came about midway between the original " evaporator " of three sections, the length of which was nearly the same, and the modified " evaporator " of 5 sections, which

* The specified dimensions were unfortunately not adhered to in constructing the still-head of thirteen sections, with the result that there was a tendency for two or three of the sections to become blocked with condensed liquid. The efficiency of the apparatus was thus increased, but its usefulness was greatly diminished and the quantity of liquid in the still-head was much greater than it should have been.

was a little longer. But the weight of liquid in the 3-section "evaporator" still-head was considerably less than half, and in 5-section "evaporator" not much more than half as great as in the Hempel tube.

Length for length, the modified "evaporator" still-head would be more efficient, and would contain only about half the amount of condensed liquid.

REFERENCES.

1. Young, "On the Relative Efficiency and Usefulness of various Forms of Still-head for Fractional Distillation, with a Description of some New Forms possessing Special Advantages," *Trans. Chem. Soc.*, 1899, **75**, 679.
2. Brown, "The Comparative Value of Different Methods of Fractional Distillation," *Trans. Chem. Soc.*, 1880, **37**, 49.
3. Wurtz, "Memoir on Butyl Alcohol," *Ann. Chim. Phys.*, 1854, III, **42**, 129.
4. Kreis, "Comparative Investigations on the Methods of Fractional Distillation," *Liebig's Annalen*, 1884, **224**, 259.
5. Hempel, "Apparatus for Fractional Distillation," *Fresenius' Zeitschr. für Anal. Chem.*, 1882, **20**, 502.

CHAPTER XI

DEPHLEGMATORS

IN many of the still-heads employed on the large scale, for example the Coffey still (Fig. 58, p. 202) the condensed liquid is made, by means of suitable obstructions, to collect into shallow pools, and the ascending vapour has to force its way through these pools; very good contact is thus brought about at definite intervals between vapour and liquid. The excess of liquid is carried back from pool to pool, and finally to the still by suitable reflux tubes. It is convenient to reserve the term "dephlegmator" for this particular class of still-head.

The "Linnemann" Dephlegmator.—The first dephlegmator employed in the laboratory was devised by Linnemann (1). A number of cups of platinum gauze, A, were placed at different heights in the vertical tube (Fig. 40) and the liquid collected in these, but, as no reflux tubes were provided, the liquid gradually accumulated in the still-head until the quantity became unmanageable, when the distillation had to be discontinued until the liquid flowed back to the still. There was thus much waste of time, and increased loss of material by evaporation, and it was impossible to make an accurate record of the temperature.

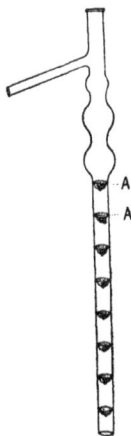

FIG. 40. — The "Linnemann" dephlegmator.

In the more recent forms of dephlegmator, reflux tubes are provided, and it is on the size and arrangement of these tubes that the efficiency and usefulness of the still-heads chiefly depends.

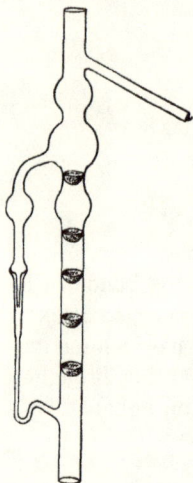

The "Glinsky" Dephlegmator.—

The Glinsky dephlegmator (2) has only one reflux tube, which carries the excess of liquid from the large bulb to the tube below the lowest obstruction, practically back to the still. In its original form (Fig. 41), the dephlegmator otherwise resembled the Linnemann still-head very closely, but, as now constructed, there are bulbs on the vertical tube and spherical glass beads, instead of platinum cups, rest on the constrictions between the bulbs.

FIG. 41.—The "Glinsky" dephlegmator.

The "Le Bel-Henninger" Dephlegmator.—

In the Le Bel-Henninger apparatus (3) the obstruction is usually caused by placing platinum cones on the constrictions between bulbs blown on the vertical tube; each bulb is connected by a reflux tube with the one below it (Fig. 42) so that the liquid is carried back from bulb to bulb and not straight to the still.

In these dephlegmators, unlike the Coffey still, the reflux tubes are external, and the returning liquid is thus exposed to the cooling action of the air. The dephlegmators of Brown (4) and of Young and Thomas (5), (6),

FIG. 42.—The "Le Bel-Henninger" dephlegmator.

follow the principle of the Coffey still more closely, the reflux tubes being much shorter and being heated by the ascending vapour.

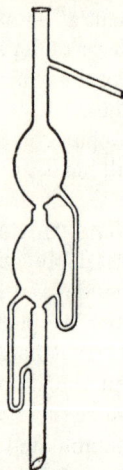

The "Young and Thomas" Dephlegmator.—The

Young and Thomas dephlegmator (6) is shown in Fig. 43 *a*, *b*, and *c*. It consists of a long, glass tube of about 17 mm. internal diameter, with the usual narrow side delivery tube. In the wide tube are sharp constrictions, on which rest concave rings of platinum gauze, R, previously softened by being heated to redness, and these support small, glass reflux tubes, T, of the form shown in Fig. 43 *b*. The upper and wider part of the reflux tube has an internal diameter of 4·5 mm., the narrow **U**-shaped part, which serves as a trap, an internal diameter of 3 mm.; if, however, the number of constrictions exceeds 10 or 12, the traps

Fig. 43.—The " Young and Thomas " dephlegmator.

for the lower reflux tubes should be slightly wider, say, 3·5 mm. The length of the reflux tubes should be about 45 mm., and the distance between two constrictions about 60 to 65 mm. The enlargement, A, on the reflux tube prevents it from slipping through the ring if the tube is inverted, and the reflux tube and ring together are prevented from falling out of position by the five internal projections (made by heating the glass with a fine blow-pipe flame and pressing it inwards with a carbon pencil) one of which, B, is shown in Fig. 43 *b*. A horizontal section through the tube at B is shown in Fig. 43 *c*.

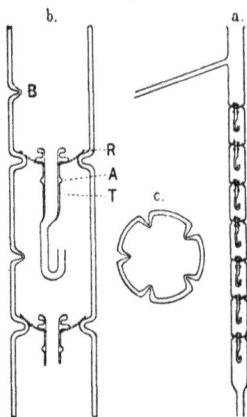

Comparison of Dephlegmators. (6)—On comparing

the efficiency of the three dephlegmators—each of three sections—it was found that when a large quantity (400 grams) of the mixture of benzene and toluene was distilled, the Le Bel-Henninger still-head gave slightly better results

than the Young and Thomas, and both of these distinctly
better results than the Glinsky. On distilling 50 grams
of the mixture, the Young and Thomas dephlegmator was
found to be the best; with the Le Bel-Henninger tube the
residual toluene was not quite pure, and with the Glinsky
it contained a quite appreciable amount of benzene, although
the temperature reached 110·6° in all three cases.

With 25 grams of the mixture the differences in efficiency
were much accentuated, and it was only with the Young
and Thomas dephlegmator that the temperature reached
110·6°, though even in this case the toluene was not quite
pure. The highest temperature reached with the Glinsky
still-head was 107·6°, and with the Le Bel-Henninger,
107·35°, but the residual toluene was far less pure in the
former case, for, on distillation from a small bulb, the Glinsky
residue came over between 102·2° and 110·4°, and the Le
Bel-Henninger from 105·7° to 110·6°, and the Young and
Thomas from 110·4° to 110·6°.

With the "rod and disc," the "pear" and the "evaporator" (3 and 5 sections) still-heads, nearly as good results
were obtained with 25 as with 50 or more grams of the
mixture, and in all these cases the residual toluene was
quite pure.

It will thus be seen that, for small quantities of liquid, the
Young and Thomas dephlegmator gives better results than
the Glinsky or Le Bel-Henninger, but that none of them are
so satisfactory as the other forms of still-head.

The relative efficiency of the three dephlegmators when
400 and 25 grams, respectively, of the mixture were distilled, and the effect of increasing the number of sections
in the case of the Young and Thomas still-head are shown
in Table 50. The results in the last column were obtained
by taking the distillation at only half the usual rate. In
this case, 42·6 out of the 50 grams of toluene were recovered
in a pure state in the single distillation, and 33·7 grams of
benzene were obtained with a temperature range of only 0·5°.

TABLE 50.

Nature of still-head.	Glinsky.	Le Bel-Henninger.	Young and Thomas.	Glinsky.	Le Bel-Henninger.	Young and Thomas.	Young and Thomas.				
Number of sections.			3			3	3	6	12	18	18
Vertical height in cm.	30	43	51*	30	43	51*	51*	78*	122*	130	130
Weight of mixture distilled.		400			25				100		
Final temperature.											
Percentage weight of distillate.											
80·7											33·7
81·2	1·2	0·6	0·3	0·6	0·4	0·6	0·4	0·6	21·1	32·4	7·2
83·2	19·1	25·8	20·3	13·2	13·4	18·6	20·2	22·8	18·6	9·6	3·4
86·2	14·2	12·1	13·9	15·0	15·2	15·2	14·5	14·5	4·3	3·0	2·0
89·2	8·7	9·2	10·4	8·8	10·0	7·6	8·6	6·3	1·9	1·5	1·2
92·3	6·6	4·3	4·5	6·8	7·4	6·0	5·2	3·1	1·8	1·2	0·9
95·4	5·5	3·1	4·2	6·0	5·0	4·6	4·5	2·5	1·3	0·9	0·7
98·5	6·1	4·3	3·3	6·4	5·2	3·8	3·7	2·1	0·8	0·8	0·3
101·6	3·7	2·9	3·4	5·2	9·6	4·6	3·9	1·7	0·8	1·0	0·4
104·6	6·2	4·8	3·8	20·8		5·6	4·9	2·0	1·0	1·1	0·6
107·6	7·1	6·0	8·3			9·8	8·5	3·9	1·8	1·4	1·2
110·0	10·0	10·4	10·5			7·6	9·3	4·6	4·1	2·8	2·8
110·6								8·0	5·5	6·0	3·0
Pure toluene by difference.	11·6	16·5	17·1	(17·2)	(28·6)	16·0	16·3	27·9	37·0	38·3	42·6
	100·0	100·0	100·0	100·0	100·0	100·0	100·0	100·0	100·0	100·0	100·0
Weight of liquid and vapour in still-head.	2·8	5·85	2·8	2·8	5·85	2·8	2·8	5·3	10·6	12·1	12·1

* These were unnecessarily long.

General Remarks on the Construction of Dephlegmators.—The results of the experiments which have been made serve to indicate the requirements which should be fulfilled in order that a dephlegmator may give the best possible results.

1. **Number of Sections.**—It should be possible to greatly increase the number of sections without seriously adding to the difficulty of construction or to the fragility of the apparatus. This requirement is best fulfilled by the "Brown" and the "Young and Thomas" dephlegmators.

2. **Size of Constrictions, &c.**—As the amount of condensed liquid flowing back at any level is greatest at the bottom of a still-head and least at the top, it follows that, in order to retard the flow sufficiently for a pool to be formed, more complete obstruction is necessary at the top of the tube than at the bottom. In the Le Bel-Henninger dephlegmator, however, the constrictions are frequently made widest, and the platinum cones largest, at the top of the tube. There is thus a tendency, on the one hand, for the liquid to flow past the upper cones without forming a pool, and, on the other hand, for the quantity of liquid in the pools in the lower bulbs to be unnecessarily large. In the Young and Thomas apparatus it is advisable to make the upper constrictions somewhat deeper than the lower ones.

3. **Width of Reflux Tubes and Depth of Traps.**—It has been pointed out that, consistently with efficiency, the amount of condensed liquid in the still-head during distillation should be as small as possible. The reflux tubes should, therefore, not be made wider—in that part which is filled with liquid during the distillation—than is necessary to freely carry back the condensed liquid ; also the U-shaped parts, acting as traps, should be no deeper than is required to prevent the ascending vapour from forcing its way through them. Additional width or depth simply means waste

liquid in the still-head, but if the number of sections is very large, the lower traps should be made rather wider than the upper ones. On the other hand, the upper part of the reflux tubes may be advantageously made fairly wide so as to facilitate the entrance of the condensed liquid, and also to prevent bubbles or columns of vapour from being caught and carried down with the liquid through the traps. Such columns of vapour, when formed, are liable to drive out the liquid, and the ascending vapour may then pass more easily through the traps than through the pools formed by the obstructions.

The poor results obtained with the Le Bel-Henninger dephlegmator, when only 25 grams of the mixture was distilled, were probably largely due to the excessive width and depth of the reflux tubes. The weight of liquid and vapour in the still-head was more than twice as great as in those of Glinsky or Young and Thomas.

4. **Flow of Liquid through the Reflux Tubes.**—It is of the utmost importance that there should be a rapid flow of condensed liquid through the reflux tubes, especially if they are outside the main tube and are not heated by the ascending vapour. To take an extreme case, suppose that there were no back flow at all through the reflux tubes and that the traps simply became filled with the first portions of condensed liquid. This most volatile liquid would thus remain lodged in the traps until the end of the distillation, and would then form part of the residue. In the distillation, for instance, of a mixture of benzene and toluene, the last fraction might consist of pure toluene, while the residue at the end of the fractionation would be very rich in benzene.

Of the three dephlegmators compared, the Young and Thomas is the best in this respect ; in the Glinsky apparatus, on the other hand, the flow of liquid was exceedingly slow, and it is for this reason that, when only 25 grams of

N

the mixture were distilled, the residue was so much richer in benzene than the last portions of distillate. The Glinsky apparatus is, indeed, quite unsuited for the distillation of very small quantities of liquid. The same fault is to be observed with the Le Bel-Henninger dephlegmator, but it is not nearly so marked.

In fractionating tubes of the Glinsky or Le Bel-Henninger type, the upper end of the reflux tube should be wide and the junction with the bulb should be low down, in order that the quantity of liquid in the pool may not become unnecessarily large. This is a point that is frequently overlooked.

5. **Arrangement of the Reflux Tubes.**—To get the best results, there should be a reflux tube connecting each section with the one below it, so that the change in composition may be regular from bottom to top of the dephlegmator. That is the case with the dephlegmators of Le Bel-Henninger, Brown, and Young and Thomas, but in the Glinsky apparatus there is only one reflux tube connecting the top bulb practically with the still. Thus the condensed liquid, which is returned to the still through the reflux tube, is richer in the more volatile component of the mixture than the liquid in the lower pools.

6. **Return of Liquid from Still-Head to Still after the Distillation is Completed.**—When the residual liquid is valuable, it is of importance that it should return as completely as possible from the still-head to the still. The weights of liquid actually left in the still-heads after cooling were as follows:—Glinsky, 0·2 gram; Le Bel-Henninger, 1·4 grams; Young and Thomas, 0·55 gram. In this respect, and this only, the Glinsky dephlegmator gave the best results, but the amounts of liquid left in the "Rod and Disc," the "Pear," and the "Evaporator" tubes were still smaller.

Of the various still-heads that have been described, it

may be concluded that, when only moderate efficiency is required, the "Rod and Disc" or "Pear" is to be most strongly recommended, but for great efficiency the "evaporator" still-heads give the best results.

Comparison of Improved with Plain Vertical Still-Head.

—The relative efficiency of the different still-heads is well shown by the following comparison of the number of fractional distillations with a plain vertical tube of 30 cm. height, which give the same result as a single distillation with the improved apparatus.

	Description of still-head.	No. of fractionations.
1.	"Rod and Disc" (20 discs)	More than 2.
2.	"Pear" (13 bulbs)	Nearly 3.
3.	"Hempel" (200 large beads)	Nearly 4.
4a.	"Evaporator," original form (3 sections) . .	More than 3.
4b.	,, ,, ,, (5 ,,) . .	About 5.
5a.	,, modified ,, (3 ,,) . .	More than 3.
5b.	,, ,, ,, (5 ,,) . .	Nearly 5.
5c.	,, ,, ,, (8 ,,) .	About 6.
6a.	Young and Thomas Dephlegmator (3 sections)	Between 2 and 3.
6b.	,, ,, ,, ,, (6 ,,)	About 4.
6c.	,, ,, ,, ,, (12 ,,)	About 7.
6d.	,, ,, ,, ,, (18 ,,)	Nearly 8.
6d.	,, ,, ,, ,, (18 ,,) } (half rate)	Nearly 9.

In general, the improvement in the separation of toluene is better than in that of benzene.

REFERENCES.

1. Linnemann, "On a Substantial Improvement in the Methods of Fractional Distillation," *Liebig's Annalen*, 1871, **160**, 195.

2. Glinsky, "An Improved Apparatus for Fractional Distillation," *Liebig's Annalen*, 1875, **175**, 381.

3. Le Bel and Henninger, "On Improved Apparatus for Fractional Distillation," *Berl. Berichte*, 1874, **7**, 1084.

4. Brown, "The Comparative Value of Different Methods of Fractional Distillation," *Trans. Chem. Soc.*, 1880, **37**, 49.

5. Young and Thomas, "A Dephlegmator for Fractional Distillation in the Laboratory," *Chem. News*, 1895, **71**, 117.

6. Young, "The Relative Efficiency and Usefulness of Various Forms of Still-Head for Fractional Distillation," *Trans. Chem. Soc.*, 1899, **75**, 679.

CHAPTER XII

MODIFICATIONS OF THE STILL-HEAD (*continued*)

"REGULATED" OR "CONSTANT TEMPERATURE" STILL-HEADS

By surrounding the still-head with water or any other liquid, the temperature of which is kept as little above the boiling point of the more volatile component as will allow of vapour passing through, a considerable improvement in the separation is effected. The temperature of the bath, however, requires very careful regulation if the boiling points of the components are near together, or if one component is present in large excess, for, in either case, a fall of a fraction of a degree would cause complete condensation of the vapour, while a rise of temperature to a similar extent would prevent any condensation from taking place and there would be no fractionation at all.

When a regulated temperature still-head is employed, it is better, for two reasons, to bend the tube into the form of a spiral; in the first place, the effective length of the still-head may be thereby greatly increased without unduly adding to the height of the bath, and in the second place, as has been already pointed out, the spiral form is more efficient than the vertical.

Warren's Still-Head.—The employment of an elongated spiral still-head, kept at a constant or slowly rising temperature, was first recommended by Warren (1). The spiral tube was heated in a bath of water or oil; its length varied from $1\frac{1}{2}$ to 10 feet and its internal diameter from $\frac{1}{4}$ to $\frac{1}{2}$ inch.

Warren carried out fractional distillations of petroleum and other complex mixtures, and observed that, as the fractions became purer, the temperature of the bath had to be brought nearer to the boiling point of the liquid in the still and required more careful regulation.

Brown's Still-Head.—A modification of this apparatus, devised with a view to the better control of the temperature of the still-head, is described by Brown (2). The liquid to be distilled is boiled in the vessel A (Fig. 44) ; the vapour rises through the coil C, the temperature of which is that of a liquid boiling in the vessel E. The vapour from the liquid in E passes through the tube D to the worm condenser F, and the condensed liquid returns by the tube K to the bottom of the vessel E. The pressure under which the jacketing liquid boils is regulated and measured by a pump and gauge connected with the tube M. The liquid in E is heated by the ring burner B.

The vapour of the liquid which is being distilled

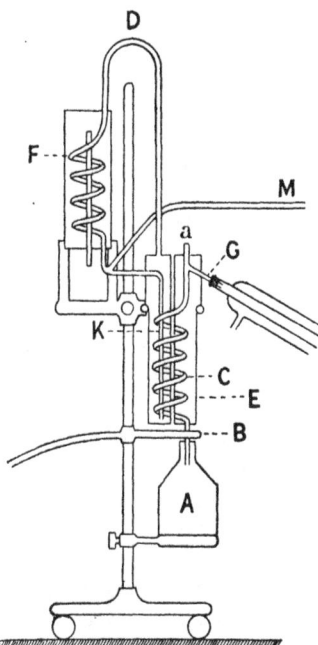

FIG. 44.—Brown's "regulated temperature" still-head.

passes through the side delivery tube G and is condensed and collected in the usual manner. The temperature of the vapour as it leaves the still-head is registered by a thermometer at a ; that of the jacketing liquid is regulated by the pressure and may be read from the vapour pressure curve, or a second thermometer may be placed in the central tube.

Brown obtained very good results with his apparatus, and arrived at the important conclusion (p. 86) that " in distillations with a still-head maintained at a constant temperature, the composition of the distillate is constant, and is identical with that of the vapour evolved by a mixture whose boiling point equals the temperature of the still-head."

In Brown's apparatus the temperature of the still-head can be kept constant for any length of time or it can be altered rapidly and easily by altering the pressure under which the jacketing liquid is boiling, but, owing perhaps to the fact that the number of observations and the amount of attention required during a distillation are greater than when the still-head is merely exposed to the cooling action of the air, it has not come into general use.

Suitability for very Volatile Liquids.—For very volatile liquids there can be no question that the regulated temperature still-head is the most suitable, and the principle has been applied by Ramsay and Dewar to the purification of gases. Thus, if a mixture of helium with any other gases is passed at the ordinary pressure through a spiral tube cooled by liquid hydrogen boiling under reduced pressure, all other known gases would be condensed in the tube and the helium alone would pass through.

Separation of Pentanes from Petroleum.—A regulated temperature still-head, combined with a dephlegmator (Fig. 45) has been found very useful for the separation of the lower paraffins from American petroleum (3). The vapour from the boiling " petroleum ether " passed first through a six column Young and Thomas dephlegmator, on leaving which its temperature was read on a thermometer, A. It then passed upwards through a spiral tube in a large bath, the water in which was either cooled by adding ice, or warmed

by a ring burner below, and was kept constantly stirred by
an arrangement similar to that employed by Ostwald, in
which a propeller, with four blades of thin sheet copper,
was kept rotating by a windmill actuated by a ring of gas

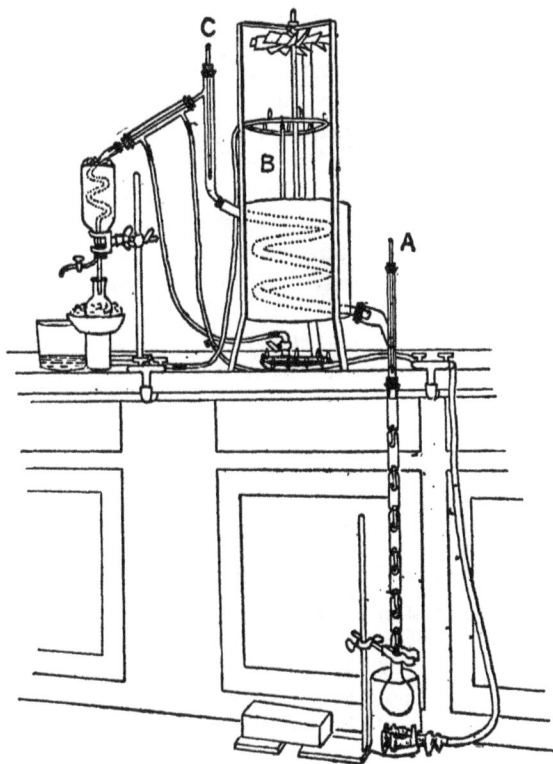

Fig. 45.—Combined "regulated temperature" still-head and "Young and Thomas'
dephlegmator.

jets, the efficiency of the windmill being greatly increased by
the use of a chimney. The temperature of the bath was
registered by the thermometer B.

After leaving this part of the apparatus the vapour passed

through a vertical tube, where its temperature was read on the thermometer c, then into a spiral condenser cooled by ice ; the condensed liquid was collected in ice-cooled flasks.

The general course of this separation has already been referred to (p. 138), and is indicated by the curves in Fig. 29. Details of a single fractionation, the eleventh, are given in Table 51; the temperatures are the final ones for the fractions and they are all corrected to 760 mm., and for the thermometric errors. Since the temperature of the vapour is affected by changes of the barometric pressure, that of the water bath has been altered in each case to the same extent. The lowest fractions consist chiefly of isopentane ; the middle ones are mixtures of iso- and normal pentane ; the highest consist of nearly pure normal pentane.

TABLE 51.

ELEVENTH FRACTIONATION.

Fraction.	D	B	T	$D - T$	ΔT	ΔW	$\dfrac{\Delta W}{\Delta T}$
1	28·42°	28·25°	28·30°	0·12°	(1·15°)	78	(70)
2	29·15	28·85	28·90	0·25	0·60	101	168
3	30·17	29·65	29·65	0·52	0·75	58	77
4	31·75	30·55	30·55	1·20	0·90	44	49
5	33·10	32·15	32·20	0·90	1·65	43	26
6	34·60	33·90	33·85	0·75	1·65	48	29
7	35·35	34·65	34·75	0·60	0·90	37	41
8	35·65	35·20	35·40	0·25	0·65	40	61
9	35·90	35·80	35·85	0·05	0·45	43	95
10	36·11	36·00	36·10	0·01	0·25	80	320
11	36·25	36·30	36·23	0·02	0·13	81	623
12	36·32	36·30	36·31	0·01	0·08	71	890

D = temperature of vapour on leaving the dephlegmator.
B = temperature of bath.
T = temperature of vapour before entering the condenser.

The bath was kept at such a temperature, and the source of heat was so regulated, that the drops of distillate fell as nearly as possible at the rate of 60 per minute, and it will be noticed that the temperature of the vapour before condensation was in all cases nearly the same at that of the bath, generally very slightly higher. It will also be seen that the fall in temperature of the vapour, during its passage through the regulated temperature still-head $(D - T)$, is greatest for the middle and least pure fractions and smallest for the highest and purest; indeed, for the last fraction the three temperatures are almost identical, while for the fourth there is a difference of 1·2° between D and T.

Very careful regulation of the temperature of the bath was necessary when the liquid distilled was nearly pure.

REFERENCES.

1. Warren, "On the Employment of Fractional Condensation," *Liebig's Annalen*, 1865, Suppl., **4**, 51.
2. Brown, "The Comparative Value of Different Methods of Fractional Distillation," *Trans. Chem. Soc.*, 1880, **37**, 49; "Fractional Distillation with a Still-Head of Uniform Temperature," *ibid.*, 1881, **39**, 517.
3. Young and Thomas, "Some Hydrocarbons from American Petroleum. I. Normal and Iso-pentane," *Trans. Chem. Soc.*, 1897, **71**, 440.

CHAPTER XIII

CONTINUOUS DISTILLATION

Continuous Distillation on Large Scale.—Continuous distillation of wort has long been carried out on the large scale by means of the Coffey still (p. 202) but it is only quite recently that any attempt has been made to devise a process of the kind suitable for laboratory purposes.

Carveth's Apparatus.—Carveth (1) suggests that by maintaining two parts of a system at different temperatures, corresponding to the boiling points of the two components of a mixture, it should be possible to effect continuous separation, especially if use were made of dephlegmating intercepts. He describes an apparatus used by Derby, in 1900, for mixtures of alcohol and water. It consists (Fig 46) of a long block-tin tube surrounded at its lower end, A, by the vapour of boiling water, at its upper end, B, by vapour from boiling alcohol, and filled with intercepts. The mixture of alcohol and water to be distilled was slowly dropped in at c, the vapour passing to the condenser at D, and the residue through the trap at E. Carveth, however, gives no details regarding the length or diameter of the tubes or the nature of the intercepts.

So far as the lower half of the still-head is concerned, there is no objection to the use of steam as a jacket, the object aimed at being to keep the temperature up to the boiling point of water in order that the alcohol may be vaporised as completely as possible; the great amount of heat evolved by the condensation of steam makes that

substance a very efficient heating agent. But the use of
the *vapour* of alcohol for the upper part of the still-head is
wrong in principle, and it is a mistake to suppose that this
part would thus be maintained " at a temperature corresponding to the boiling point of alcohol." What is here
required is to keep the temperature as far as possible from
rising above the boiling point of alcohol, in order that the
vapour of water may be condensed as completely as possible,
while that of the alcohol passes on to the condenser. This
can be done by surrounding the still-head with a *liquid*
which is kept at the required temperature by suitable means,

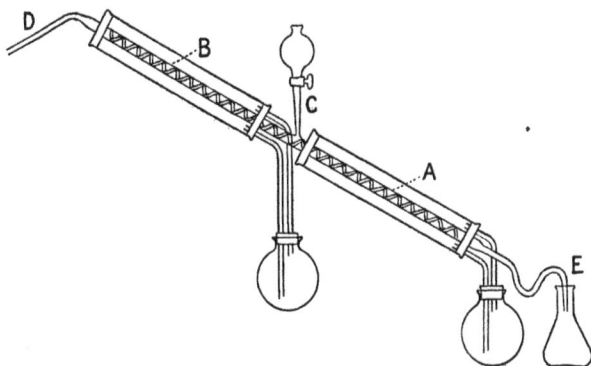

FIG. 46.—Carveth's still for continuous distillation.

but if a vapour, such as that of the more volatile component,
is used as a jacket, it easily becomes superheated and the
still-head is not prevented from rising in temperature at all ;
indeed the effect is very similar to that produced by covering the tube with cotton-wool—condensation is somewhat
diminished and so also is the efficiency.

Carveth makes the general statement that the percentage
of alcohol by weight in the residue was found to average
about 0·5 when working carefully, but in five cases for
which he gives details the percentage varied from 1·5 to 2·0.
These five distillations were carried out with extreme

slowness, the average rate being only 7·5 grams of distillate per hour.

Again, the percentage of alcohol in the distillate is stated to have varied from 90·6 to 93·9 in four cases, but in the five very slow distillations referred to above the percentages varied from 76·7 to 89·9. It does not appear possible to form any very definite idea of the efficiency of the apparatus from the data given.

Lord Rayleigh's Apparatus.—Lord Rayleigh (2) describes an apparatus similar in principle to Carveth's except that the temperatures of both parts of the still-head are regulated by liquids and not by vapours.

The apparatus consists of a long length (12 metres) of copper tubing, 15 mm. in diameter, arranged in two spirals which are mounted in separate pails. For the distillation of mixtures of alcohol and water the lower and longer spiral was heated by boiling water, the upper one by water maintained at a suitable temperature, usually 77°. The spirals were connected by a straight glass or brass tube of somewhat greater bore, provided with a lateral junction through which the mixture could be introduced. With the exception of the two extremities, the whole length of tubing sloped gently and uniformly upwards from near the bottom of the lower pail to the top of the upper pail, where it turned downwards and was connected with an ordinary Liebig's condenser. The lower end of the tubing was, if necessary, connected with an air-tight receiver heated to 100°.

The mixture was introduced at such a rate that it fell in a rapid but visible succession of drops and, when rich in alcohol, it was previously heated. Mixtures containing 20, 40, 60 and 75 per cent. of alcohol, were distilled and in all cases the water, collected in the lower receiver, was nearly pure, never containing more than 0·5 per cent. of alcohol. The distillate varied but little in strength and contained from 89 to 90·3 per cent. of alcohol.

It seems probable that such a continuous process may prove very useful for mixtures which separate into only two components and when the quantity to be distilled is large.

Continuous Separation of Three Components.—It would, moreover, be possible to devise an arrangement by which three components could be separated in a similar manner, but it would be necessary to have two coils for the middle substance; the form of apparatus required is shown

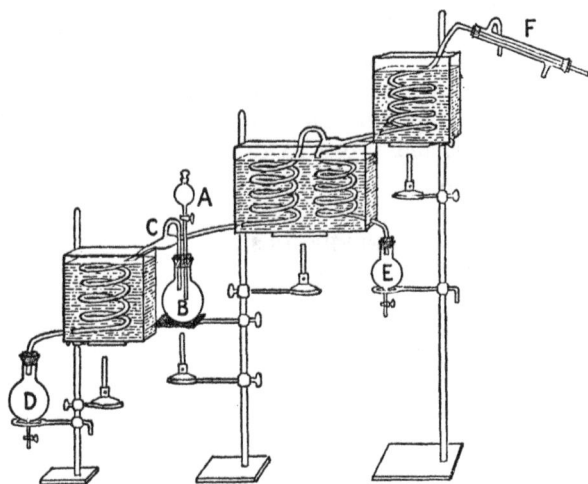

Fig. 47.—Still for continuous distillation of three components. (Adapted from Lord Rayleigh's still.)

diagrammatically in Fig. 47. Suppose, for example, that the liquids to be separated were methyl, ethyl and propyl acetates (b. p. 57·1°, 77·15° and 101·55° respectively). The highest bath would be kept at 57·1°, the middle one at 77·15° and the lowest at 101·55°. The mixture would be introduced slowly through the funnel at A, into the still, B, where it would be boiled, and the mixed vapour would enter the still-head at C. The propyl acetate would be collected in the receiver D, which might be heated by a ring burner.

The vapour in passing upwards through the first coil in the middle bath would be freed, more or less completely, from propyl acetate and the mixture of ethyl and methyl acetates would enter the top of the second coil, chiefly in the form of vapour. The condensed liquid which reached the bottom of the coil, and was collected in E, would be nearly pure ethyl acetate; while the vapour that reached the top of the coil in the highest bath, and was condensed in F, would be nearly pure methyl acetate.

It should thus be theoretically possible to separate all three liquids in a pure state, and the components of a still more complex mixture should be separable, if there were as many baths as components kept at the boiling points of those components, and two coils in each bath except the lowest and highest.

REFERENCES.

1. Carveth, " Studies in Vapour Composition," Part II, *Journ. Phys. Chem.*, 1902, **6**, 253.
2. Rayleigh, "On the Distillation of Binary Mixtures," *Phil. Mag.*, 1902, [VI], **4**, 521.

CHAPTER XIV

Benzene and Toluene.—That much time is saved by the use of an improved still-head is seen by comparing the results of a fractional distillation of 200 grams of a mixture of equal weights of benzene and toluene with a modified "evaporator" still-head of 5 sections with those obtained with a plain vertical tube 30 cm. in height.

When the "evaporator" still-head was used it was not necessary to divide the distillate into more than five fractions. In the first distillation (Table 52), more than half the total quantity of toluene was recovered in a pure state, and it will be noticed that the values of $\Delta w/\Delta t$ for the middle fractions are relatively very low.

In the second fractionation, a little liquid came over at the boiling point of benzene, but it was not considered advisable to collect this separately. The value of $\Delta w/\Delta t$ for the first fraction was more than 300 times greater than for the fourth. More than 82 per cent. of the toluene was now obtained in a pure state.

In the third fractionation 21·9 grams of pure benzene were obtained, and when fraction No. 3, collected from 81·4° to 95·4°, was added to the residue in the still, and the distillation was continued, the temperature did not rise above 81·7°; the whole of the distillate was therefore collected in receiver No. 2, and the small residue (4·95 grams) was rejected. The apparatus was then dried, and the higher fractions were distilled, when only 7·25 grams came over below 110·6°, the residue (9·3 grams) consisting of pure toluene. The distillate was too small to allow of further fractionation.

Two more distillations of the lower fractions sufficed to complete the recovery of the benzene.

TABLE 52.

	I.			II.		
No. of fraction.	Temperature range.	Δw	$\dfrac{\Delta w}{\Delta t}$	Temperature range.	Δw	$\dfrac{\Delta w}{\Delta t}$
1	...	...	...	80·2— 80·4°	39·25	**196·2**
2	81·5— 83 2°	51·6	**30·4**	80·4— 81·4	43·55	**43·5**
3	83·2— 95·4	47·1	**3·9**	81·4— 95·4	16·5	**1·2**
4	95·4—110·0	30·1	**2·1**	95·4—110·0	8·9	**0·6**
5	110·0—110·6	19·8	**33·0**	110·0—110·6	7·9	**13·2**
Toluene .	110·6°	51·0	∞	110·6°	82·65	∞
		199·6			198·75	

	III.		IV.		V.	
No. of fraction.	Temperature range.	Δw	Temperature range.	Δw	Temperature range.	Δw
Benzene	80·2°	21·9	80·2°	56·3	80·2°	81·8
1	80·2—80·3°	44·35	80·2—80·25°	30·7		
2	80·3–81·7	27·0				
Residue	...	4·95		10·3		15·0
		98·2		97·3		96·8
...	Rejected below 110·6°	} 7·25				
Toluene	110·6°	91·95				
		197·4				

Comparative Results with Plain and Improved Still-Heads.

—The great improvement effected by employing an efficient apparatus is clearly shown in Table 53.

TABLE 53.

	Evaporator still-head of 5 sections.	Plain still-head 30 cm. in height.
Weight of pure benzene recovered .	81·8 grams	. . 81·4 grams
„ „ toluene „ .	91·95 „	. . 88·8 „
Time required for actual distillation.	6½ hours	About 30 hours.
Loss by evaporation, &c.	4·0 grams	. . 22·0 grams.
Mixture left undistilled	22·25 „	. . 7·8 „

The chief gain was in the time occupied, which was reduced nearly to one-fifth, but the recovery also was somewhat better, and in many cases would be much better ; there was also much less actual loss of material, though the amount left undistilled was greater.

Methyl, Ethyl and Propyl Acetate.

—The separation of a mixture of 100 grams of methyl acetate, 120 grams of ethyl acetate and 100 grams of propyl acetate was carried out by distillation with a modified evaporator still-head of 8 sections. Details of the fractionations are given in Table 54 (p. 194).

In the second fractionation, when fraction No. 7 from I. had been distilled over as completely as possible, the temperature had risen only to 81·5°. Fraction 8 from I. was, however, added to the residue in the still as usual and, on distillation, a considerable amount was collected below 89·35°, but the quantity that came over between this temperature and 100° was very small, and it was clearly not worth while to attempt to recover any more ethyl acetate in the next fractionation from the eighth fraction of II. This last fraction was therefore merely redistilled to recover as much propyl acetate as possible, and all that came over below 101·55 (5·15 grams) was rejected.

Similarly, in the third fractionation, when fraction 3 from II. had been distilled as completely as possible, the temperature had only risen to 65°, but fraction 4 from II.

o

TABLE 54.

	I.			II.			III.		
No. of fraction	Temperature range.	Δw	$\dfrac{\Delta w}{\Delta t}$	Temperature range.	Δw	$\dfrac{\Delta w}{\Delta t}$	Temperature range.	Δw	$\dfrac{\Delta w}{\Delta t}$
1	...	...	...	57·1 — 57·4°	26·25	87·5	Methyl acetate	9·45	∞
							57·1 —57·25°	42·2	281·3
2	Below 60·85°	49·65	?	57·4 — 59·4	49·15	24·6	57·25—58·05	30·5	38·1
3	60·85— 67·1	48·3	7·7	59·4 — 67·1	24·1	3·3	58·05—67·1	15·45	1·7
4	...	...	...	67·1 — 76·15	15·75	1·7	67·1 —76·65	11·3	1·2
5	67·1 — 77·15	55·85	5·6	76·15 - 77·15	43·4	3·9	76·65—77·15	41·0	82·0
6	...	...	...	77·15— 77·65	45·65	91·3	77·15—77·3	46·9	312·7
7	77·15— 89·35	64·85	5·3	77·65— 89·35	13·45	1·1	77·3 —77·7	12·65	31·6
8	89·35—101·35	34·45	2·9	89·35—101·55	14·2	1·2	Residue . .	6·3	
9	101·35—101·55	11·45	57·2					215·75	
							Rejected below 101·55°	5·15	
	Propyl acetate	54·65	∞	Propyl acetate	85·35	∞	Propyl acetate	91·2	∞
		319·2			317·3			315·1	

	IV.		V.		VI.	
No. of fraction	Temperature range.	Δw	Temperature range.	Δw	Temperature range.	Δw
	Methyl acetate .	34·45	Methyl acetate . .	58·9	Methyl acetate . .	76·8
1	57·1—57·2° . . .	32·0	57·1—57·15° ' . .	24·0	Total residue . .	18·15
2	57·2—58·0 . . .	23·1	Total residue . .	12·55		94·95
	Residue	6·55		95·45		
		96·1				
	Rejected below 74·0°	3·35				
4	74·0—77·1° . . .	12·7	Rejected below 76·5°	5·5		
5	77·1—77·15 . . .	36·05	76·5—77·15° . . .	10·6	Rejected below 77·15°	8·2
6	Ethyl acetate . .	52·55	Ethyl acetate . .	88·1	Ethyl acetate . .	95·8
		200·75		199·65		198·95
	Total residue .	12·6				
		213·35				

was added, and the distillation continued as usual. In fractionation IV., after fraction 3 from III. had been added to the residue and the distillation continued, the quantity of liquid left in the still was extremely small when the temperature reached 58°; the distillation was therefore stopped, and the residue was rejected. The fractionations numbered III. to VI. were therefore not continuous, but in each case consisted of two parts. Thus in III. the distillation was stopped when the temperature had reached 77·7° ; the still and still-head were dried, and the last fraction from II. was separately distilled, when the recovery of propyl acetate was completed.

In IV. the first part of the fractionation ended when the temperature reached 58°, and was recommenced by the distillation of fraction 4 from III. In this case the temperature did not rise above 77·15°, even when the highest fraction from III. was distilled ; fraction 6 was therefore taken to be pure ethyl acetate.

Comparative Results with Plain and Improved Still-Head.— The great superiority of the "evaporator" still-head is clearly shown by the results given in Table 55.

TABLE 55.

	Long plain still-head.	"Evaporator" still-head of 8 sections.
Percentage weight of methyl acetate recovered	48·1	76·8
,, ,, ethyl acetate ,,	53·5	79·8
,, ,, propyl acetate ,,	72·5	94·2
Number of fractionations required	32	6
Time required in hours	About 70	17
Percentage weight of material—		
(1) Lost by evaporation and transference	21·0	2·8
(2) Lost by treatment with reagents	5·6	0·0
(3) Left undistilled	17·1	13·8

In the long continued fractionation with the plain still-head, a considerable amount of moisture was absorbed by the esters and some hydrolysis took place. Before the fractionations were completed it was necessary to treat the propyl acetate with potassium carbonate and the methyl acetate with phosphorus pentoxide, and the loss was thus increased.

With the "evaporator" apparatus, on the other hand, the propyl acetate showed no acid reaction whatever, and the methyl acetate was only slightly moist.

The specific gravities were again determined and were found to agree well with those given on p. 136.

Fractional Distillation under Reduced Pressure.—The improved still-heads may be employed for distillation under reduced pressure. A 12-column Young and Thomas dephlegmator, for example, was used by Francis (1) for the separation of isoheptyl and normal heptyl bromides from the products formed by the action of bromine on the distillate from American petroleum coming over between 93·5° and 102°. The pressure in this case was 70 mm.

REFERENCE.

1. Francis and Young, "Separation of Normal Isoheptane from American Petroleum," *Trans. Chem. Soc.*, 1898, **73**, 920.

CHAPTER XV

DISTILLATION ON THE MANUFACTURING SCALE

The Still.—The still is usually made of copper or some other metal, but for special purposes other materials must be employed. Thus, for the distillation of sulphuric acid, vessels of glass or platinum are used; for nitric acid, cast fron is found to be the best material; stoneware, or glass, stills are employed for bromine and iodine, and so on.

THE STILL-HEAD.

When only a rough separation of the constituents of a complex mixture is required, as in the preliminary distillation of coal tar, dephlegmators or rectifiers are not used, but the vapours pass directly from the still to the condenser.

For the better separation of the components of a mixture, modified still-heads are employed.

Mansfield's Still.—A simple contrivance, which is often used, is to cool the still-head with water, the temperature of which is allowed to rise to a suitable extent; this device is adopted in Mansfield's still (Fig. 48). The head, A, is surrounded by water which becomes heated to its boiling point. Liquids which boil at temperatures higher than 100° are for the most part condensed in the still-head and return to the still; afterwards the stopcock, B, is opened and the vapour then passes directly to the condenser. With this still, Mansfield was able to separate benzene in a fairly pure state from coal tar.

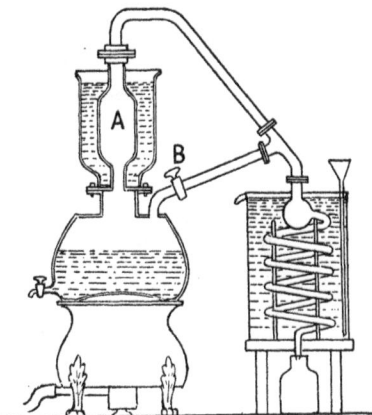

Fig. 48.—Mansfield's still.

Coupier's Still.—In Coupier's still (Fig. 49), a more elaborate cooling arrangement is combined with a dephlegmator. The vapour from the still, A, passes first through the rectifier or dephlegmator, B, and then by the pipe, C, into a series of bulbs placed in a cistern, D, containing brine, which may be warmed by steam from the pipe, E. The liquid condensed in the bulbs returns by the pipes, F, F, to the dephlegmator, the less volatile vapours being for the most part condensed in the first bulb and the liquid returning to a low part of the dephlegmator, while that from the other bulbs reaches it successively at higher levels. The vapour,

freed from substances boiling at higher temperatures than
that of the water in the tank, then passes by the pipe, G, to

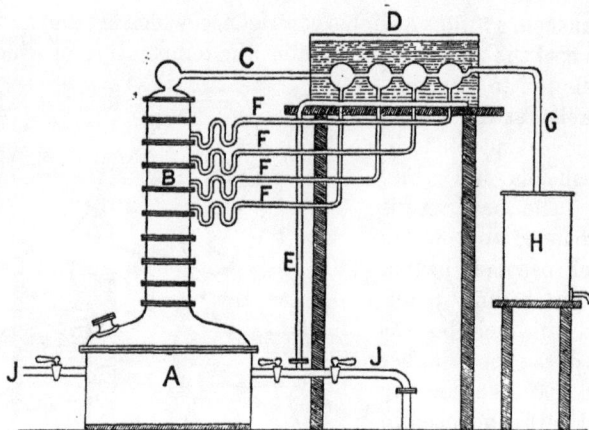

FIG. 49.—Coupier's still.

the condenser, H. The contents of the still are heated by
the steam-pipe, J.

French Column Apparatus.—A very similar arrangement is
seen in the French column apparatus (Fig. 50). The liquid
to be distilled is heated by a steam-pipe in the still, A; the
vapour rises through the rectifier, B, passes through the
series of pipes in the tank, C, and then enters the condenser,
D. Cold water from the cistern, E, enters the condenser at
the bottom, and in its passage upwards is warmed by the
condensation of the vapour; the warm water passes through
the pipe, F, into the tank, which is divided into sections by
the vertical partitions shown in the figure. Here the water
receives more heat owing to condensation of vapour in the
pipes, so that it is hottest where it leaves the tank by the
pipe, G. Thus the vapour, in passing from the rectifier to
the condenser, is cooled by successive stages, and the liquid

which returns to the lower part of the rectifier through the
pipe, H, is much richer in the less volatile components than
that which reaches the top
by the pipe H'.

Savalle's Still.—In the
Savalle apparatus for the
separation of benzene,
toluene and xylene from
crude naphtha, the va-
pours pass first through a
dephlegmator or analysing
column, then through a
surface or multitubular
condenser which is pro-
vided with a water supply
so regulated that its temperature is about that of
the boiling point of the

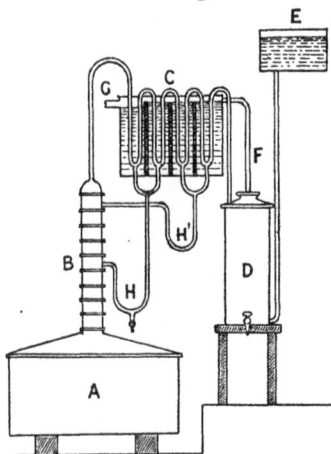

Fig. 50.—French column apparatus.

liquid required ; the temperatures in this case would be
about 80° for benzene, 110° for toluene, and 140° for the
xylenes. The liquid condensed in the regulated temperature
still-head returns as usual to the dephlegmator, and the
purified vapour passes on to the cold condenser.

Dephlegmators.—In nearly all the dephlegmators, rectifiers, or analysers, part of the vapour is condensed and forms
a pool of liquid through which the vapour that follows has to
force its way, so that very complete contact between vapour
and liquid is ensured. The details of the various dephlegmators differ, however, considerably, as will be seen from
Figs. 51 to 55.

Dubrunfaut's Dephlegmator.—Dubrunfaut's method is
shown in Fig. 51. The vapour rises through the central
pipe, A, in the tray, but its ascent is barred by the dome, B,
and it bubbles through the condensed liquid in the tray.

The excess of liquid flows back through the pipe, c, the lower end of which is trapped by the liquid in the tray below. The construction of the dome is indicated in the lowest tray in the figure. In a large column a number of these domes are arranged in a ring on each tray.

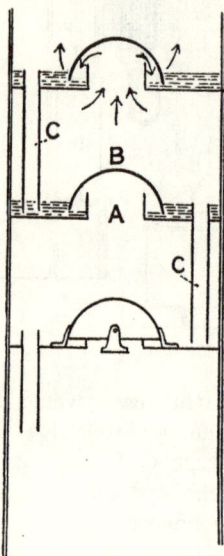

a

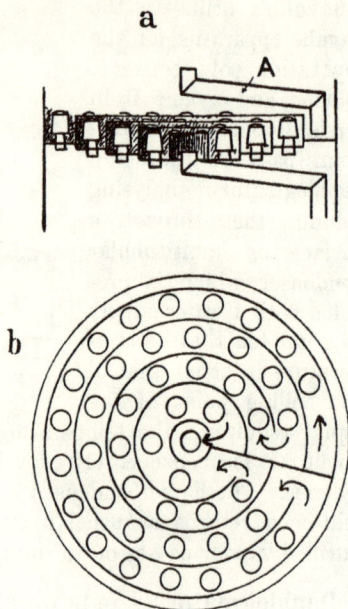

Fig. 51.—Dubrunfaut's dephlegmator.   Fig. 52.—Egrot's dephlegmator.

Egrot's Dephlegmator.—In the Egrot dephlegmator (Fig. 52), an arrangement somewhat similar to the dome is used, but the condensed liquid is made to follow a zigzag course from the circumference to the centre of each tray. The liquid enters the tray by the pipe A, shown in Fig. 52 a, and follows the course shown by the arrows in Fig. 52 b.

Savalle's and Coffey's Dephlegmators.—In many cases the trays are perforated, the vapour rises through the perforations and bubbles through the liquid, the excess of which descends through the reflux tubes. Figs. 53 and 54 show

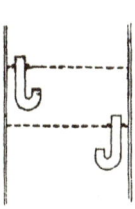

FIG. 53.—Savalle's dephlegmator.

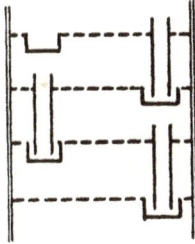

FIG. 54.—Savalle's dephlegmator.

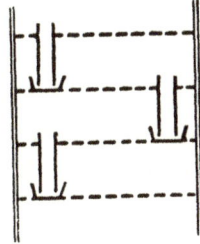

FIG. 55.—Coffey's dephlegmator.

the arrangement adopted in two of Savalle's stills ; that in the Coffey still, which is described on p. 202, is shown in Fig. 55.

The "Pistorius" Still.—A quite different form of still-head is employed in the Pistorius still (Fig. 56). Here the vapour entering a section of the still-head is deflected from the centre to the circumference by the flat dome, A ; it then

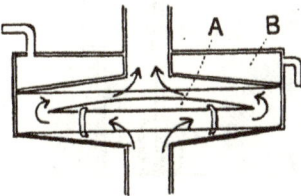

FIG. 56.—The "Pistorius" still.

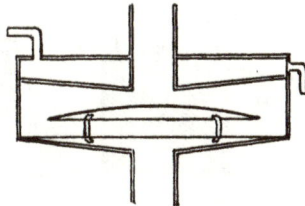

FIG. 57.—Modified Pistorius still.

passes back to the centre above the dome and is partially condensed by the water in B, above it. It is probable that a considerable improvement in the efficiency of this apparatus would be effected by the simple modification shown in Fig. 57. The condensed liquid, instead of flowing down the outer walls of the section would drop from the central tube on to the flat dome as in the "evaporator" still-head (p. 166) and the ascending vapour would come into better contact with this hot liquid and would cause some of it to evaporate again.

CONTINUOUS DISTILLATION.

The "Coffey" Still.—Various alterations in the still have been made in order to make the process of distillation a continuous one. In this country very strong spirit is

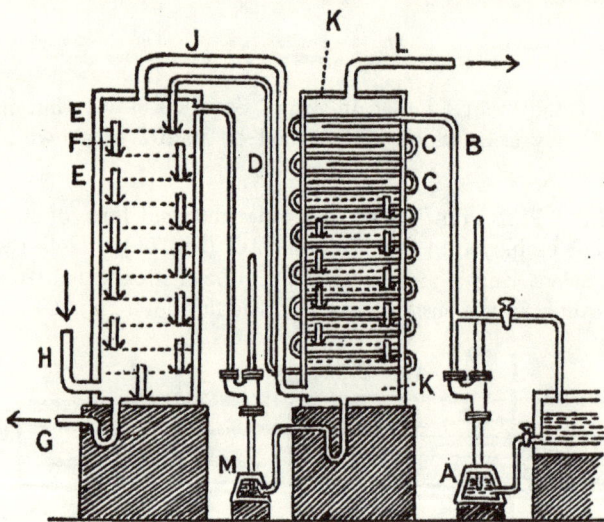

Fig. 58.—The "Coffey" still.

obtained in a single operation from wort by means of the Coffey still, the essential parts of which are shown in Fig. 58. The wort is pumped from a reservoir, A, up the pipe, B, and passes down the zigzag pipe, C C, where it is heated by the ascending vapour ; then up the pipe, D, into the highest section of the dephlegmator or analyser, E. It

then descends from section to section through the tubes,
F, and is finally allowed to escape through the trapped
pipe, G.

Steam is passed into the analyser by the pipe, H, and
causes the wort to boil, so that by the time it has reached
the bottom it is completely deprived of alcohol. The
ascending vapours pass through the perforations in the
plates and bubble through the liquid on them, a portion of
the aqueous vapour being thus condensed, and the descending wort heated, by each washing.

On reaching the top of the column, the concentrated
alcoholic vapour passes through the pipe J into the bottom
of the rectifier, K, and then ascends through perforated
plates similar to those in the analyser ; the ascending
vapour is, however, not washed by wort in the rectifier but
by the weak spirit formed by partial condensation of the
vapour. In the upper part of the rectifier there are usually
only shelves which compel the vapour to take the same
zigzag course as the pipes which convey the wort downwards.
The purified vapour, containing about 91 per cent. by weight
of alcohol, then passes through the pipe L to the condenser.

The weak spirit condensed in the rectifier flows into a
reservoir, M, from which it is pumped into the top of the
analyser, where it mixes with the descending wort.

REFERENCES.

1. Thorpe, Articles on "Distillation," "Benzene," "Alcohol,"
   Dictionary of Applied Chemistry.
2. Payen, "Alcohol," *Précis de Chimie Industrielle*, Vol. II

CHAPTER XVI

FRACTIONAL DISTILLATION AS A METHOD OF QUANTITATIVE ANALYSIS

Determination of Composition of Mixture.—The composition of a mixture of liquids which are not difficult to separate may, as a rule, be ascertained with a fair degree of accuracy from the results of a single distillation with an efficient still-head, or, if the components are more difficult to separate, from the results of two or three distillations.

It will be well to consider first the case of mixtures which tend, on distillation, to separate normally into the original components.

Taking first the simplest case, that of a mixture of two liquids, it is found that the weight of distillate that comes over below the 'middle point' is, as a rule, almost exactly equal to that of the more volatile component, even when the separation is very far from complete.

By 'middle point' is to be understood in all cases the temperature midway between the boiling points of the two components, whether single substances or mixtures of constant boiling point, into which the original mixture tends to separate; or, in the case of more complex mixtures, the temperature midway between the boiling points of any two consecutive fractions of constant boiling point.

If the original mixture contains more than two, say n, components, which separate normally on distillation, the weights of these components will be very nearly equal respectively to (No. 1) the weight of distillate below the first middle point, (No. 2 to $n-1$) the weight of distillate between the successive middle points, (No. n) the weight above the last middle point.

Loss by Evaporation.—It is obvious that there must be some loss by evaporation, which always makes the weight of distillate somewhat too low. This loss will be greater as the initial boiling point of the liquid is lower and as the temperature of the room is higher. It is not, however, proportional to the amount of liquid distilled, for a great part of it is caused by the saturation of the air in the flask and still-head with vapour when the liquid is first heated; since this vapour is mixed with much air its partial pressure is low, and a large proportion of it escapes condensation when cooled in the condenser. Under otherwise similar conditions the loss is, therefore, roughly proportional to the volume of air in the still and still-head, and it is advantageous to use as small a flask as possible and to employ a still-head of as small a capacity as is consistent with efficiency.

Choice of Still-Head.—A plain wide tube or one with spherical bulbs is the least satisfactory, but the " pear " still-head, owing to the diminished capacity of the bulbs and the increased efficiency, gives much better results. Of all forms, the " evaporator " is the best, because the capacity is very small relatively to the efficiency, and the amount of condensed liquid in it is smaller than in any other equally efficient still-head; moreover, almost the whole of the condensed liquid returns to the still at the end of the distillation. With a liquid of low viscosity, like one of the lower paraffins, the quantity left in the still head is almost inappreciable, and in other cases it may be reduced to a very small amount by disconnecting the apparatus, while hot, from the condenser, tilting the tube from side to side to facilitate the flow of liquid back to the still and, if the original form of " evaporator " still-head is used, shaking out any liquid remaining in the funnels.

Estimation of Loss by Evaporation. —The following may be taken as an example of the estimation of loss by evaporation. Mixtures of benzene and methyl alcohol, one with benzene, the other with methyl alcohol in excess (these liquids form a mixture of minimum boiling point), were distilled through an " evaporator " still-head of five sections of the original form, the distillation being stopped in each case when the middle point was reached. The following results were obtained :

TABLE 56.

	Component in excess.	
	Benzene.	Methyl alcohol.
Weight of distillate	128·7	132·0
Weight of liquid in still	24·9	27·2
Total	153·6	159·2
Weight of mixture taken	154·2	160·1
Loss by evaporation and left in still-head	0·6	0·9

When the benzene was in excess, it is certain that the amount of it left in the still-head did not exceed 0·1 gram, and the loss by evaporation was therefore estimated as 0·5 gram ; in calculating the composition this amount was added to the observed weight of distillate. With methyl alcohol in excess the total loss was greater, but this more viscous liquid does not flow back so completely to the still, and the loss by evaporation was taken to be the same, 0·5 gram. With an " evaporator " apparatus of five sections the loss by evaporation is usually from 0·3 to 0·5 gram.

Mixtures Containing Two Components.—The following are examples of the distillation of mixtures of two liquids which separate normally into the original components.

TABLE 57.

METHYL ALCOHOL AND WATER.

Boiling points—Methyl alcohol, 64·7° ; water, 100° ; middle
point, 82·35°.

I.—*Methyl alcohol in large excess.*

Mixture taken.	Weight of distillate below middle point.	Percentage composition of mixture.		
		Found.		Taken.
		Uncorrected.	Corrected.	
Alcohol 90·9	Observed 90·5	Alcohol 78·5	**78·7**	**78·8**
Water 24·4	Corrected 90·8	Water 21·5	**21·3**	**21·2**
115·3		100·0	**100·0**	**100·0**

II.—*Water in large excess.*

Alcohol 39·7	Observed 33·9	Alcohol 16·9	**17·0**	**19·7**
Water 161·5	Corrected 34·2	Water 83·1	**83·0**	**80·3**
201·2		100·0	**100·0**	**100·0**

The first result is quite satisfactory, while the second is
not, but it must be remembered that it is always difficult to
separate the more volatile component of a mixture when
present in relatively small quantity, and, in such a case,
a second distillation is usually necessary. The first distillation was therefore continued until the temperature
reached 100°, and the whole of the distillate, weighing 66·8
grams, was then redistilled and the double correction for loss
by evaporation was applied.

The weight below the middle point was now 38·9, corrected 39·5, giving the percentage composition.

	Uncorrected.		Corrected.		Taken.
Alcohol	19·3	. .	**19·6**	. .	**19·7**
Water	80·7	. .	**80·4**	. .	**80·3**
	100·0		**100·0**		**100·0**

It will thus be seen that, by repeating the distillation, the result was as satisfactory as that given by a single distillation when the alcohol was in excess. Even without correcting for loss by evaporation the agreement is fairly good, but it is much improved by introducing the correction.

TABLE 58.

ISOAMYL ALCOHOL AND BENZENE.

Boiling points—Benzene, 80·2° ; isoamyl alcohol, 132·05° ; middle point, 106·1°.

Mixture taken.	Weight below middle point.	Percentage composition of mixture.		
		Found.		Taken.
		Uncorrected.	Corrected.	
Alcohol 26·6	Observed 85·55	Alcohol 23·8	**23·6**	**23·7**
Benzene 85·7	Corrected 85·85	Benzene 76·2	**76·4**	**76·3**
112·3		100·0	**100·0**	**100·0**

Here the agreement is very satisfactory.

One Component in Large Excess.—That a single distillation may be sufficient when the more volatile of two components is present in large excess, while two or more distillations are necessary when it is present in relatively small amount, is further shown by the following results.

A mixture containing 90 grams of benzene and 10 grams of toluene was distilled through an evaporator still-head of three sections.

Weight below middle point : observed,   89·6 ; corrected, 89·9.
Percentage of benzene in mixture : taken, **90·0** ; found,   **89·9.**

When 100 grams of a mixture containing only 10 per cent. of benzene was distilled through the same still-head, very little came over below the middle point, 95·4°, and the quantity was too small to admit of fractional distillation. By twice redistilling all that came over below 110·6°, however, the weight below the middle point rose to 9·0 grams. Allowing a loss of 0·3 gram for each distillation the corrected weight would be 9·9 instead 10 grams.

With a larger quantity, 250 grams, a fairly satisfactory result was obtained even with a " pear " still-head. In this case the distillate was divided into three fractions, and the following results were obtained.

TABLE 59.

	I.	II.	III.	IV.	V.
1. Below 95·4° . . .	0	16·2	21·1	22·7	23·2
2. 95·4—104·7° . . .	43·7	21·0	10·9	6·4	3·5
3. 104·7—110·5 . . .	76·0	39·4	19·4	9·6	6·3
	119·7	76·6	51·4	38·7	33·0

The weight below the middle point is clearly approaching a limit, the increase being smaller each time ; allowing 0·3 gram for loss by evaporation in each distillation the last weight would be 23·2 + 1·5 = 24·7, and the percentage 9·9 instead of 10.

Advantages of Efficient Still-Head.—A great saving of time is, however, effected, and a more certain result is obtained by the use of a very efficient still-head. Thus, on distilling 300 grams of the above mixture through an 18-column Young and Thomas dephlegmator, the first distillation gave 21·4 grams below the middle point, and a total of 76·1 below the boiling point of toluene. On redistillation of the 76·1 grams the weight below the middle point was 29·2 grams, so that with no correction for loss by evaporation the calculated percentage of benzene would be 9·7 and, with the larger still-head it would be fair to allow 0·4 gram for each distillation, which would bring up the weight to 30·0 grams, and the percentage of benzene to 10·0.

Mixtures of Three Components.—The fractionation of a mixture of methyl, ethyl and propyl acetates with a plain

P

vertical still-head one metre in length has already been referred to (p. 128), and some details have been given. With so large numbers of fractions and of fractional distillations, it would not be possible to distribute the loss by evaporation, and by transference from receiver to still, between the different fractions, but the total loss was ascertained in each fractionation, and we may calculate the percentages on the total quantity of material left at the end of each operation instead of on the original quantity taken.

The boiling points of the three esters are 57·1°, 77·15°, and 101·55° respectively, and the two middle points are therefore 67·1° and 89·35°. The percentage weights of distillate below 67·1°, from 67·1° to 89·35°, and above 89·35° were read from the curves (Fig. 28) and are given below:

TABLE 60.

Number of fractionation.	Percentage weight of distillate.		
	Below 67·1°.	From 67·1° to 89·35°.	Above 89·35°.
1	11·5	74·5	14·0
2	22·5	45·5	32·0
3	28·5	41·5	30·0
4	32·5	36·5	31·0
5	31·0	38·5	30·5
6	30·0	39·5	30·5
7	31·5	38·5	30·0
8	30·0	39·0	31·0
9	29·0	40·0	31·0
10	31·0	39·0	30·0
11	31·0	37·7	31·3
12	30·5	38·3	31·2
Mean of last 9 percentages .	30·7	38·6	30·7
Percentage taken .	31·7	38·2	30·1

It will be seen that the numbers remain nearly constant
after the first three fractionations, and that the mean percentages calculated from the last nine distillations agree
fairly well with those in the original mixture. The actual
loss by evaporation must have been greatest for methyl
acetate, and least for propyl acetate and the calculated
percentages are too low for the first and too high for the
second of these esters.

Advantages of Efficient Still-Head.—With an evaporator
still-head of eight sections the following results were obtained (p. 193).

Weight of methyl acetate taken	100
,, ethyl ,, ,, ...	120
,, propyl ,, ,, ...	100
	320

Weight of distillate below first middle point, 97·95 ; corrected, 98·45.
Weight of distillate between first and second
middle points . . . . 120·7 ;   ,,   120·8

Percentage composition of mixture.

	Found.	Taken.
Methyl acetate	30·77	31·25
Ethyl ,, ...	37·75	37·50
Propyl ,, ...	31·48	31·25
	100·00	100·00

The distillation was continued until the temperature
reached the boiling point of propyl acetate. The residue in
the still was weighed when cold, and the loss, due partly to
evaporation and partly to the minute amount of liquid left
in the still-head was found to be 0·8 gram. It was assumed
that of this loss, 0·7 gram was due to evaporation, and that
0·5 gram was lost below the first middle point, 0·1 between
the two middle points, and 0·1 above the second middle
point.

The fractionation of the mixture was continued, and the

results obtained in the second complete distillation are given below.

Weight below first middle point	99·5
Weight between first and second middle points	118·25
Weight above second middle point	99·55
	317·3
Total loss	2·7
	320·0

Here, again, it is impossible to distribute the loss correctly between the different fractions, and it is best to calculate the percentages on the total amount of material left at the end of the distillation.

Percentage composition of mixture.

	Found.	Taken.
Methyl acetate	31·36	31·25
Ethyl ,,	37·27	37·50
Propyl ,,	31·37	31·25
	100·00	100·00

The agreement is better than after a single distillation, and is very satisfactory.

Complex Mixtures.—The separation of isopentane and normal pentane (b. p. 27·95° and 36·3°; middle point 32·15°) from a mixture containing also butanes, hexanes and a very little pentamethylene has been already referred to (p. 138). Taking the weight of distillate between 27·95° and 36·3° as 100 in each case, the percentage coming over between 27·95° and 32·15° became roughly constant after the first distillation. The variation in this case was greater than with the esters (40—46, mean 42 per cent. in 12 fractionations), but the difference between the boiling points of the components is only 8·35° against 20·15° and 22·4°.

It would appear, then, that in American petroleum, of the two paraffins, about 42 per cent. consists of isopentane and 58 per cent. of normal pentane. At the end of the fractionations 101 grams of pure isopentane and 175 grams of pure normal pentane were obtained, or 36·6 per cent. of the isoparaffin ; but the loss by evaporation must have been greater, and there is never so good a recovery of the more volatile component.

Mixtures of Constant Boiling Point.—For the sake of brevity, a mixture of constant boiling point containing two components will be referred to in this chapter simply as a " binary " mixture, and a mixture of constant boiling point containing three components as a " ternary " mixture.

The quantity of a mixture of constant boiling point may be estimated by the distillation method in exactly the same way as that of a single substance. The methods of experiment and of calculation are similar in all respects.

Two examples may be given to prove this point ; binary mixtures of isopropyl alcohol and tertiary butyl alcohol respectively with water were mixed with excess of water and distilled with the following results.

TABLE 61.

ISOPROPYL ALCOHOL AND WATER.

Boiling points : Binary mixture, 80·37° ; water, 100·0° ; middle point, 90·2°.

Mixture taken.	Weight below middle point.	Percentage composition of mixture.	
		Found.	Taken.
Binary mixture 57·7	Observed 57·3	Binary mixture 74·05	74·15
Water . . . . 20·1	Corrected 57·6	Water . . . . 25·95	25·85
77·8		100·00	100·00

TABLE 62.

TERTIARY BUTYL ALCOHOL AND WATER.

Boiling points : Binary mixture, 79·9° ; water, 100·0° ; middle
point, 89·95°.

Mixture taken.	Weight below middle point.	Percentage composition of mixture.	
		Found.	Taken.
Binary mixture 58·8	Observed 58·2	Binary mixture **66·25**	**66·6**
Water . . . . 29·5	Corrected 58·5	Water . . . . **33·75**	**33·4**
88·3		**100·00**	**100·0**

The found percentages have been calculated in both cases
from the corrected weights of distillate. The agreement
is very good in the first case and satisfactory in the second.

DETERMINATION OF THE COMPOSITION OF MIXTURES OF
CONSTANT BOILING POINT BY DISTILLATION

Binary Mixtures.—Since a mixture of constant boiling
point behaves like a single substance on distillation, it is
possible, if we know the composition of the mixture distilled,
to calculate that of the binary mixture.

For a mixture of minimum boiling point, the ratio of the
weight of the component not in excess in the original
mixture to the corrected weight of distillate below the
middle point is equal to the proportion of that component
in the binary mixture.

In the case of a mixture of maximum boiling point, the
ratio of the weight of the component not in excess to that of
the residue after the middle point has been reached is equal
to the proportion of that component in the binary mixture

The following examples may be taken.

TABLE 63.

I. *Normal propyl alcohol and water, with the latter in excess.*

Boiling points: Binary mixture, 87·72° ; water, 100·0° ; middle
point, 93·85°.

Mixture taken.	Weight below middle point.	Percentage composition of binary mixture.	
		Distillation method.	From specific gravity.
Alcohol . . . 76·6	Observed 106·4	Alcohol 71·8	71·69
Water . . . 50·0	Corrected 106·7	Water 28·2	28·31
126·6		100·0	100·00

The calculation is carried out as follows.

Weight of propyl alcohol, 76·6 grams.

Weight of binary mixture = corrected weight of distillate = 106·7 grams.

$$\text{Percentage of alcohol in binary mixture} = \frac{76 \cdot 6 \times 100}{106 \cdot 7} = 71 \cdot 8.$$

In the calculation of the composition from the specific
gravity of the redistilled binary mixture, the necessary
correction has been introduced for the contraction that
occurs on mixing the components.   In many cases the boiling
point of the binary mixture is too near that of one of the
components to allow of a determination of composition
being made with that component in excess, but if the boiling
point is greatly depressed one may frequently determine the
composition even when the more volatile of the two original
components is in excess.

Thus with methyl alcohol and benzene two separate determinations were made with the following results.

TABLE 64.

I. *Benzene in excess.*

Boiling points : Binary mixture, 58·34° ; benzene, 80·2° ; middle
point, 69·25°.

II. *Methyl alcohol in excess.*

Boiling points : Binary mixture, 58·34° ; methyl alcohol, 64·7° ;
middle point, 61·5°.

Mixture taken.		Weight below middle point.		Percentage composition of binary mixture.	
I.	II.	I.	II.	I.	II.
Alcohol 51·2	79·9	Observed 128·7	132·0	Alcohol **39·6**	**39·5**
Benzene 103·0	80·2	Corrected 129·2	132·5	Benzene **60·4**	**60·5**
154·2	160·1			**100·0**	**100·0**

Non-miscible and Partially Miscible Liquids.—The method
is applicable to liquids which are non-miscible or miscible
within limits.

Thus, with isoamyl alcohol and water, which are partially
miscible, the following results were obtained.

TABLE 65.

I. *Water in excess.*

Boiling points : Binary mixture, 95·15° ; water, 100·0° ;
middle point, 97·6°.

II. *Isoamyl alcohol in excess.*

Boiling points : Binary mixture, 95·15° ; alcohol, 132·05° ;
middle point, 113·6°.

Mixture taken.		Weight below middle point.		Percentage composition of binary mixture.	
I.	II.	I.	II.	I.	II.
Alcohol 38·8	68·3	Observed 76·4	85·65	Alcohol **50·5**	**50·3**
Water 69·5	42·7	Corrected 76·9	85·95	Water **49·5**	**49·7**
108·3	111·0			**100·0**	**100·0**

Ternary Mixtures.—When a mixture of three liquids gives rise, on distillation, to the formation of a ternary mixture of minimum boiling point, the separation may, theoretically, take place in twelve different ways and, in addition to these, if the original mixture has the same composition as the ternary mixture, its behaviour on distillation would be precisely that of a pure liquid.

Ethyl-Alcohol—Benzene—Water.—As an example we may consider mixtures of ethyl alcohol, benzene and water. For convenience, the components are represented by the initial letters A, B, and W.

The possible cases are as follows :—

	First fraction.	Second fraction.	Residue.		First fraction.	Second fraction.	Residue.
1	$A.B.W.$	$A.W.$	$W.$	7	$A.B.W.$	...	$A.$
2	,,	$B.W.$	$W.$	8	,,	...	$B.$
3	,,	$A.W.$	$A.$	9	,,	...	$W.$
4	,,	$A.B.$	$A.$	10	,,	...	$A.B.$
5	,,	$B.W.$	$B.$	11	,,	...	$A.W.$
6	,,	$A.B.$	$B.$	12	,,	...	$B.W.$
			13	Distillate $= A.B.W.$			

The first six cases and, on redistillation of the first fraction, the thirteenth, would be those commonly met with. Of the first six, the third is unrealisable in practice, owing to the very small difference between the boiling points of the second fraction ($A.W.$) and the residue (ethyl alcohol).

Mixtures, however, tending to separate in the other five ways specified, were distilled in order to determine the composition of the ternary mixture.

·· For the calculation it is necessary to know, not only the composition of the original mixture, but also that of the binary mixture forming the second fraction.

Data Required.—The boiling points of all possible components and the percentage composition of the three binary mixtures are given in Table 66.

TABLE 66.

	Boiling points.	Percentage composition.		
		A.	B.	W.
W.	100·0°	...	...	100
B.	80·2	...	100	...
A.	78·3	100	...	...
A.W. . . .	78·15	95·57	...	4·43
B.W.	69·25	...	91·17	8·83
A.B.	68·24	32·36	67·64	...
A.B.W. . .	64·86	...	...	...

The middle points are therefore as follows :—

TABLE 67.

Fractions.	Middle points.	
	First.°	Second.
I. A.B.W.; A.W.; W.	71·55°	89·1°
II. A.B.W.; B.W.; W.	67·05	84·6
IV. A.B.W.; A.B.; A.	66·55	73·3
V. A.B.W.; B.W.; B.	67·05	74·7
VI. A.B.W.; A.B. ; B.	66·55	74·2

Experimental Results.—In Table 68 are given :—(a) the actual weight of the components in the mixtures distilled, (b) the weights of distillate below and between the middle points, (c) the calculated percentages of the components in the ternary mixture.

Method of Calculation.—In calculating the composition of the ternary mixture it is assumed, as before, that the corrected weights of the two distillates are equal to those of the ternary and binary mixtures respectively which would be obtained if the separation were perfect. That being so, in case I., the weight of benzene in the ternary mixture is simply that in the original mixture; the weight of alcohol is that taken, less the amount in the binary mixture, which can be calculated; the weight of water is given by difference.

TABLE 68.

(a) Mixtures taken.

	I.	II.	IV.	V.	VI.
Alcohol	66·0	18·4	75·0	18·5	35·0
Benzene	74·2	120·0	108·0	160·1	148·3
Water	50·5	52·1	7·5	12·1	7·6
	190·7	190·5	190·5	190·7	190·9

(b) Weights below and between the middle points.

	I.	II.	IV.	V.	VI.
First.—Observed	99·5	94·9	100·6	97·1	111·6
Corrected	99·9	95·3	101·0	97·5	112·0
Second.—Observed	51·7	54·0	47·5	52·5	42·6
Corrected	51·8	54·1	47·6	52·6	42·7

(c) Percentage composition of ternary mixture.

	I.	II.	IV.	V.	VI.	Mean.
Alcohol	16·5	19·3	17·5	19·0	18·9	**18·2**
Benzene	74·3	74·2	75·1	73·4	74·3	**74·3**
Water	9·2	6·5	7·4	7·6	6·8	**7·5**
	100·0	100·0	100·0	100·0	100·0	**100·0**

The composition of the ternary mixture was also directly determined, and it will be seen that the agreement with the mean value obtained by the distillation method is very satisfactory.

	Direct determination.		Distillation method.
Alcohol	**18·5**	...	**18·2**
Benzene	**74·1**	...	**74·3**
Water	**7·4**	...	**7·5**
	100·0		**100·0**

On the other hand some of the individual values, notably those of alcohol and water in I., differ somewhat widely from the mean. The explanation of the rather large errors in this distillation is given on p. 221.

Cases to which the Distillation Method is Inapplicable.

—When small quantities of alcohol are successively added to water, the boiling point is rapidly lowered; the middle temperature between the boiling points of the pure components is, in fact, reached when the mixture contains 6·5 molecules per cent. of ethyl alcohol. On the other hand, water must be added to alcohol until the mixture contains 25 molecules per cent. before the boiling point rises 0·1° above that of pure alcohol, and with equal molecular proportions the rise of temperature is only 1·5° (1).

Very similar results are obtained with normal hexane and benzene; a mixture containing 16 molecules per cent. of benzene boils only 0·1° higher than normal hexane, and the mixture which has the boiling point 74·6°, midway between those of the pure components, contains 79 molecules per cent. of benzene (2).

In both cases mixtures of minimum boiling point, very rich in the more volatile component, are formed, so that the separation would be that of the mixture of constant boiling point from that component which is in excess.

Form of Boiling Point-Composition Curve.

—It is, however, found practically to be impossible in either case to separate the mixture of minimum boiling point even from the less volatile component, although the difference between their boiling points is considerable. In both cases the boiling point-composition curve is very flat where the more volatile component is in large excess, and it is in such cases —when the curve is very flat at either one end or the other —that one at least of the components is exceedingly difficult to separate, and that the distillation method cannot be relied on for the determination of composition.

Ethyl Alcohol and Water.—Thus on distilling ethyl alcohol-water mixtures, containing from 15 to 25 per cent. by weight of water, through an 18-column dephlegmator and calculating the percentage of water in the mixture of constant boiling point in the usual way from the weight of distillate below the middle point, values from 7·6 to 8·0 instead of 4·43 per cent. were obtained. Referring back to the calculation of the composition of the ternary ethyl alcohol-benzene-water mixture from the first distillation (p. 219), if we take 7·8 as the percentage of water in the binary *A. W.* mixture, the calculated composition of the ternary mixture becomes :

Alcohol	18·2
Benzene	74·3
Water	7·5
	100·0

which agrees very well indeed with that observed.

General Conclusions.—The distillation method may, in the great majority of cases, be safely employed for the determination of the composition of a mixture which separates normally into its components, provided that a very efficient still-head be employed and that the distillation be carried out slowly. But it must be borne in mind that from a mixture of two liquids it is almost always more difficult to separate the more volatile component than the other, and therefore, if the original mixture contain a relatively very small amount of that component, a second distillation may be necessary, and a large quantity of the original mixture will be required in order to give a sufficient amount of distillate for a second operation.

As regards the separation of three or more components from a mixture, it must be remembered that, as a general rule, the least volatile component is the easiest, while the intermediate components are the most difficult, to separate. If a binary mixture of constant boiling point is formed, the composition of the original mixture may be determined if that of the mixture of constant boiling point is known; or if the composition of the original mixture is known, that of the mixture of constant boiling point may be determined. The method may even be applied to the determination of the composition of a ternary mixture of constant boiling point.

It appears to be only when the separation of the components (either simple substances or mixtures of constant boiling point) by distillation is exceedingly difficult that the method is inapplicable.

REFERENCES.

Young, "Experiments on Fractional Distillation," *Journ. Soc. Chem. Industry*, 1900, **19**, 1072.

Young and Fortey, "Fractional Distillation as a Method of Quantitative Analysis," *Trans. Chem. Soc.*, 1902, **81**, 752.

1. Noyes and Warfel, "The Boiling Point Curves of Mixtures of Ethyl Alcohol and Water," *Journ. Amer. Chem. Soc.*, 1901, **23**, 463.

2. Jackson and Young, "Specific Gravities and Boiling Points of Mixtures of Benzene and Normal Hexane," *Trans. Chem. Soc.*, 1898, **73**, 923.

CHAPTER XVII

METHODS BY WHICH THE COMPOSITION OF MIXTURES OF CONSTANT BOILING POINT MAY BE DETERMINED

Distillation Method.—In the last chapter it has been shown how the composition of a mixture of constant boiling point may be ascertained from the weight of distillate that comes over below the middle point, when a mixture of known composition is distilled. This method is generally, but not universally applicable.

There are several other methods by which the composition of a mixture of constant boiling point may be determined.

2. By Separation of Pure Mixture.—The most accurate method—applicable, however, only to those mixtures for which the first method can be employed—is to separate the mixture of constant boiling point in a pure state by fractional distillation and to determine its composition either (a) by chemical analysis, (b) by the removal of one component, (c) from its specific gravity, or (d) from its refractive power.

(a) If one of the substances is an acid, or a base, the ordinary methods of volumetric analysis may be conveniently employed, or if one component contains a halogen, sulphur, etc., the amount of that element may be determined; but this method is not, as a rule, to be recommended.

(b) When one component is easily soluble in water, and the other insoluble, or nearly so, for example alcohol and benzene, a fairly accurate result may be obtained by shaking the mixture with water in a separating funnel and washing the insoluble component once or twice with water. The volume of this component at a known temperature may then be ascertained or its weight determined, but there is inevitably some loss by evaporation and by adhesion to the sides of the separating funnel and to the solid dehydrating agent, if this is added. As a rule, also, a little of the component that is insoluble in water remains dissolved by the aqueous solution of the other constituent, and to obtain an accurate result it would be necessary to distil this solution, and to treat the first small portion of the distillate with more water in order to separate the remainder of the insoluble component.

This method was employed for the direct determination of the composition of the ternary mixture of ethyl alcohol, benzene and water (p. 219), the benzene being determined in the manner described above and the alcohol from the specific gravity of the aqueous solution (1).

(c and d) As the specific gravities and refractive powers of mixtures are not usually strictly additive properties, it is almost always necessary to determine the values for a prepared mixture of about the same composition as that which boils at a constant temperature or, better, to determine the values for a series of mixtures in order to find what correction must be applied. Such series of determinations of specific gravity have been made by different observers in the case of mixtures of the lower alcohols with water (2) and by Brown (3) for some other pairs of liquids. The specific refractive powers have been determined for several series of mixtures by Lehfeldt (4) and by Zawidski (5).

3. **Method of Successive Approximations.**—Mixtures of different composition may be distilled, and, by successive approximations, that mixture may finally be made up which distils (a) at a constant temperature, or (b) without change of specific gravity.

(a) This method was employed by Ryland (6) to ascertain the approximate composition of the large number of mixtures of constant boiling point examined by him.

(b) If the boiling point of the mixture differs only slightly from that of either of the pure components, observations of the temperature would be useless, but we may find what mixture gives a distillate of the same specific gravity (or refractive power) as itself or, better, collecting the distillate each time in three or four fractions, we may proceed until the first and last fraction have the same specific gravity. The last method has been employed in the case of ethyl alcohol and water (2), and the following results were obtained with the two last mixtures.

TABLE 69.

I.			II.		
Weight of fraction.	Sp. gr. at 0°/4°.		Weight of fraction.	Sp. gr. at 0°/4°.	
23·6	. . .	0·81936	21·2	. . .	0·81946
73·4	. . .	...	55·0	. . .	...
27·6	. . .	0·81927	26·0	. . .	0·81953
			15·0	. . .	...
			26·1	. . .	0·81954

In the first case, the last fraction has a lower specific gravity than the first, showing that alcohol was in excess ; in the second case, it is the first fraction which has the lower specific gravity and therefore there was excess of water in the still.   It is clear that the specific gravity of the mixture that distils without change of composition must be between those (0·81936 and 0·81946) of the first fractions in these distillations.

Mapping the specific gravities as abscissæ against the weights of distillate as ordinates in each case, it is found that the lines slope almost equally, the first to the left and the second to the right (I. and II., Fig. 59), and it may therefore be assumed that the required specific gravity is 0·80941, the mean of the other two.   If the two lines are produced, they intersect each other at a point between 0·81941 and 0·81942.

According to Mendeléeff's tables, the percentage of alcohol in a mixture which has the specific gravity 0·81941 at 0°/4° is 95·57.

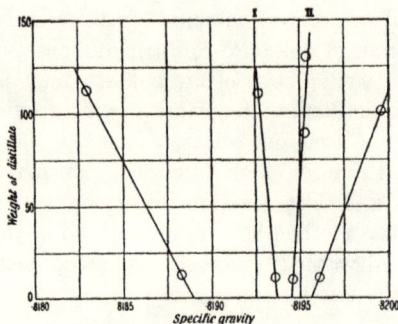

FIG. 59.—Ethyl alcohol and water.

4. Graphically from Vapour Pressures or Boiling Points.

—If the vapour pressures at constant temperature, or the boiling points under constant pressure, of a series of mixtures of known composition have been determined, these values may be mapped against the percentages of one of the components, and the percentage corresponding to the maximum or minimum pressure or temperature can then be read off. The pressure- (molecular) composition curve for carbon disulphide and methylal * is shown in Fig. 60, but it will be seen that while the maximum pressure can be read with considerable accuracy the corresponding percentage of carbon disulphide can only be roughly estimated. The same objection applies to the boiling point-composition curve.

5. Graphically from Composition of Liquid and Vapour.

—If the relative composition of vapour and liquid has been determined for a series of mixtures, the composi-

---

* The molecular weights of these two substances are equal and the molecular percentages are therefore equal to the percentages by weight.

tion of the mixture of constant boiling point may be ascertained in various ways.

(a) The percentages by weight, p, or the molecular percentages, m, of one component in the liquid may be plotted against the percentages, p' or m', of the same component in the vapour (7).

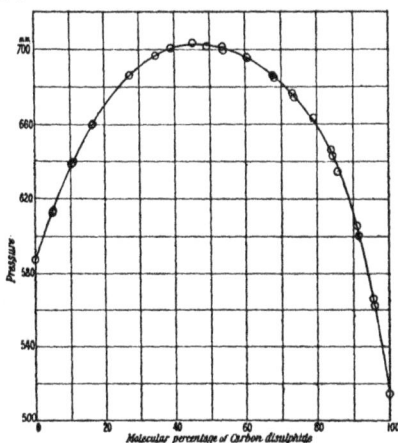

Fig. 60.—Carbon disulphide and methylal.

(b) The ratios, R, of the weights, or of the number of gram-molecules, of the two components A and B in the liquid may be mapped against the corresponding ratios, R', in the vapour.

(c) The logarithms of these ratios may be plotted in the same way (4).

Whichever method is adopted, the composition will be given by that point on the curve at which the ordinate and abscissa have the same value or, in other words, by the point of intersection of the curve with the straight line corresponding to equal values of p and p', m and m', R and R', or log R and log R'.

As an example of the first method, Lord Rayleigh's determinations of the composition of liquid and vapour for

mixtures of hydrogen chloride and water (Fig. 61) may be
mentioned (7) ; for the second and third methods we may

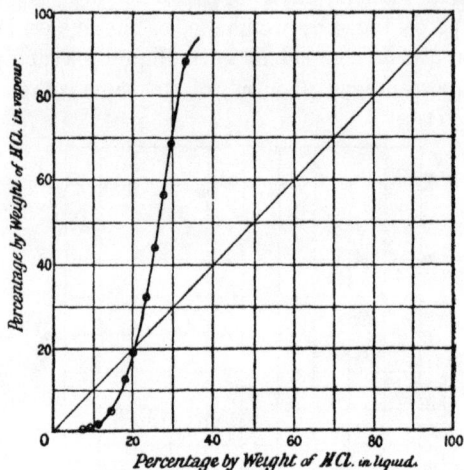

FIG. 61.—Hydrogen chloride and water.

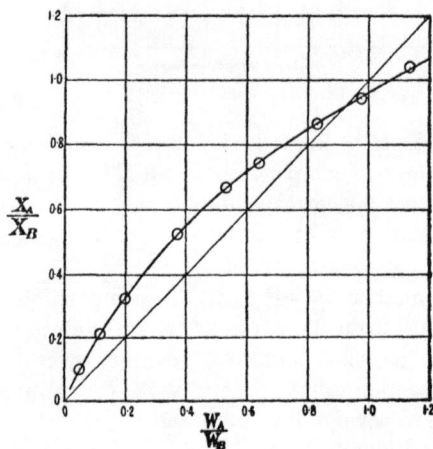

FIG. 62.—Carbon disulphide and methylal

take the results obtained by Zawidski for mixtures of carbon
disulphide and methylal (Figs. 62 and 63).

6. Graphically by means of Brown's Formula.—The relative number of molecules (or the relative weights) of the components in the liquid, W_A and W_B, and in the vapour, x_A and x_B, may be calculated from the experimental observations and the values of $\dfrac{x_B}{x_A} \cdot \dfrac{W_A}{W_B}$ plotted against the percentage number of molecules (or percentages by weight) of one component.

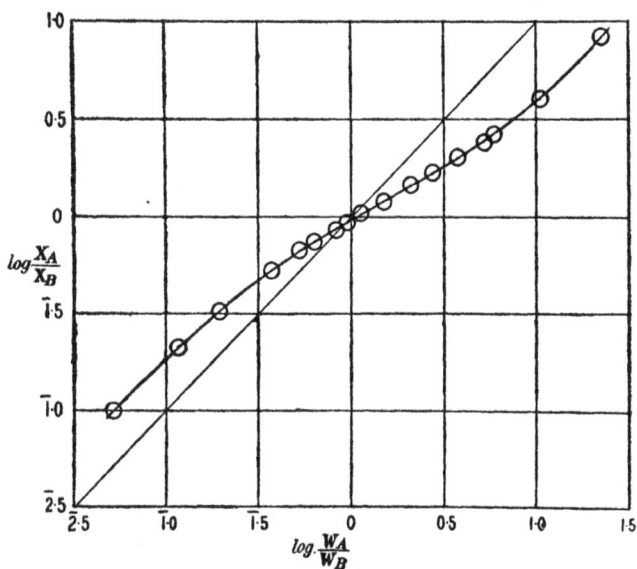

Fig. 63.—Carbon disulphide and methylal.

The percentage corresponding to $\dfrac{x_B}{x_A} \cdot \dfrac{W_A}{W_B} = 1$ is that required.

Here, again, we may make use of Zawidski's data for carbon disulphide and methylal (Fig. 64).

Results Obtained.—For benzene and ethyl alcohol the results given in Table 70 (p. 230) have been obtained.

TABLE 70.

Method.	Observer.	Percentage of benzene by weight.	Temperature.
1.	Young and Fortey	67·55	Boiling point under normal pressure.
2c.	,, ,, ...	67·64	
2b.	Ryland	68·1	
2b.	,, ...	72·1	50—51°
5c.	Lehfeldt	71·3	50°
6.	,, ...	71·3	50

When the mixture that distils without change of composition boils at almost exactly the same temperature as one

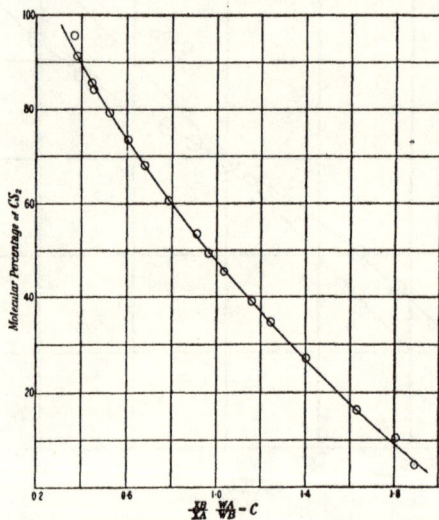

FIG. 64.—Carbon disulphide and methylal.

of the components, it is probable that methods 5, 6 and 3b are the only ones that can be relied upon, and it is on the first and second of these methods that the conclusion is based (p. 109) that benzene forms a mixture of constant boiling point with carbon tetrachloride, while the exact composition of the ethyl alcohol-water mixture has been determined by method 3b.

REFERENCES.

1. Young and Fortey, "The Properties of Mixtures of the Lower Alcohols with Benzene and with Benzene and Water," *Trans. Chem. Soc.*, 1902, **81**, 739.
2. Young and Fortey, "The Properties of Mixtures of the Lower Alcohols with Water," *Trans. Chem. Soc.*, 1902, **81**, 717.
3. F. D. Brown, "Theory of Fractional Distillation," *Trans. Chem. Soc.*, 1879, **35**, 547 ; "On the Distillation of Mixtures of Carbon Disulphide and Carbon Tetrachloride," *ibid.*, 1881, **39**, 304.
4. Lehfeldt, "On the Properties of Liquid Mixtures," Part II., *Phil. Mag.*, 1898, [V], **46**, 42.
5. Zawidski, "On the Vapour Pressures of Binary Mixtures of Liquids," *Zeitschr. physik. Chem.* 1900, **35**, 134.
6. Ryland, "Liquid Mixtures of Constant Boiling Point," *Amer. Chem. Journ.*, 1899, **22**, 384.
7. Rayleigh, "On the Distillation of Binary Mixtures," *Phil. Mag.*, 1902, [VI], **4**, 521.

CHAPTER XVIII

INDIRECT METHOD OF SEPARATING THE COMPONENTS OF A MIXTURE OF CONSTANT BOILING POINT.

Distillation after the addition of a third substance.

—It has been pointed out that there are many cases in which the two components of a mixture cannot be separated by fractional distillation, owing to the formation of a mixture of constant boiling point. It is, however, sometimes possible to eliminate one of them by adding a third substance and then distilling the mixture.

Formation of Binary Mixture of Minimum Boiling Point.

—Suppose, for example, that we have a mixture of isobutyl alcohol and benzene containing 10 per cent. by weight of the alcohol. The components cannot be separated by fractional distillation because a mixture of minimum boiling point (79·93°), containing 9·3 per cent. of isobutyl alcohol, is formed. Neither can they be satisfactorily separated by treatment with water, owing to the fact that the solubility of isobutyl alcohol in benzene is greater than in water.

If, however, a little water be added and the mixture be distilled, the first fraction will consist of the binary mixture of benzene and water, boiling at 69·25° and containing 91·2 per cent. of benzene.

In order to remove the benzene, all that is necessary is to add water to the mixture in the ratio of 8·7 to 100 parts by weight and to distil with an efficient still-head. The mixture will then tend to separate into (a) the binary benzene-water mixture (b.p. 69·25°), and (b) pure isobutyl alcohol (b.p. 108·05)°. Here the difference between the boiling points is considerable, and the separation is an easy one.

If too much water were added we should have, as an intermediate fraction, the binary alcohol-water mixture which boils at 89·8°, and contains 66·8 per cent. of the alcohol.

If, on the other hand, too little water were added, only a part of the benzene would be carried over with it and the remainder would form the binary benzene-alcohol mixture ; the residue in either case would consist of pure isobutyl alcohol.

Formation of Ternary Mixture of Minimum Boiling Point.—The substance added frequently forms a ternary mixture of constant boiling point with the two components of the original mixture, but the relative weights of these components in the ternary mixture differ from those in the binary mixture of constant boiling point which they themselves form.

Tertiary Butyl Alcohol and Water with Benzene.—Take, for example, the case of tertiary butyl alcohol and water. This alcohol is a crystalline solid, melting at 25·53° and boiling at 82·55° ; when liquefied, it is miscible with water in all pro-

portions and forms with it a mixture of constant boiling
point (79·91°) containing 88·24 per cent. of the alcohol.
The difference between the boiling point of the binary
mixture and of the alcohol is only 2·64° and the separation
of the last traces of water, or rather of the binary mixture,
from the alcohol by distillation is difficult. It was not, in
fact, found possible to obtain the alcohol quite pure by this
method, the highest melting point of the distilled alcohol
observed being 25·25°, and the boiling point 82·45°.
Fractional crystallisation gave a better result, the melting
point of the recrystallised alcohol being 25·43°.

Eventually, however, it was found that the last traces of
water could best be removed by distillation with benzene,
both the melting and boiling points (given above) of the
residual alcohol being higher than those of the product
purified by recrystallisation. Table 71 gives the boiling
points and the composition of the binary and ternary
mixtures and the boiling points of the single components.

TABLE 71.

	Boiling point.	Percentage composition.		
		Alcohol.	Benzene.	Water.
Water	100·0°	...	...	100
Tertiary butyl alcohol. . . .	82·55	100	...	..
Benzene	80·2	...	100	...
Alcohol-water	79·9	88·24	...	11·76
Alcohol-benzene	73·95	36·6	63·4	...
Benzene-water	69·25	...	91·17	8·83
Alcohol-benzene-water	67·30	21·4	70·5	8·1

It will be seen that the ratio of water to alcohol in the
binary mixture $= \dfrac{11\cdot76}{88\cdot24} = 0\cdot133$ and in the ternary mixture

$= \dfrac{8\cdot1}{21\cdot4} = 0\cdot379$, so that the latter contains nearly three times
as much water, relatively to the alcohol, as the binary mixture
does.

Not only can we remove the last traces of water from the nearly pure alcohol by means of benzene; but it is also possible to obtain the pure alcohol from the binary alcohol-water mixture.

Suppose that we start with 100 grams of this mixture; the water in it, 11·76 grams, would require $\dfrac{11·76 \times 70·5}{8·1} = 102$ grams of benzene for conversion into the ternary mixture, and if a perfect separation could be effected by a single distillation of the mixture, 145 grams of the ternary mixture and 57 grams of alcohol would be obtained. In practice, however, we should get a rather smaller amount of the ternary mixture, a little binary alcohol-benzene mixture and a residue of alcohol, the second and third fractions still containing a little water.

It is better to add a larger amount of benzene, say 125 grams, in the first place, when the quantity of the second fraction (alcohol-benzene) will be increased and the smaller amount of residual alcohol will be obtained free from water in a single operation.

Results of Distillation.—In an actual experiment, 117·5 grams of the alcohol-water mixture (containing 103·7 grams of tertiary butyl alcohol and 13·8 grams of water) with 145 grams of benzene were distilled through a 5-section "evaporator" still-head of the original form, and the following fractions were collected.

TABLE 72.

	Temperature range.	Weight.	Theoretical composition.		
			Alcohol.	Benzene.	Water.
1.	67·3—70·6° . . .	169·3	36·2	119·4	13·7
2.	70·6—78·2 . . .	39·9	14·6	25·3	
3.	78·2—82·55 . . .	18·9 ⎫	52·7		
	Residue, alcohol.	33·8 ⎭			
	Loss	0·6			
		262·5			

The residue solidified on cooling.

If the separation had been complete, the fractions would have had the composition stated in Table 72, but the first fraction must really have contained rather more alcohol and less benzene and water, and the second rather more benzene and a very little water. The third fraction must have contained some benzene with the dry alcohol.

Treatment of "Fractions."—The first fraction separated into two layers, the lower one consisting chiefly of alcohol and water with a little benzene, the upper one of benzene containing some alcohol and a little water. By adding more water in a separating funnel, running off the aqueous alcohol and washing the benzene repeatedly with small quantities of water to extract the remaining alcohol, a dilute aqueous solution of the alcohol could be obtained almost free from benzene, and by fractional distillation almost the whole of the alcohol could be recovered in the form of the binary alcohol-water mixture of constant boiling point.

Theoretically, the weight of this mixture would be 41·0 grams and to remove the water 41·8 grams of benzene would be required. But the second fraction contains about 25 grams of benzene and it would be necessary, in practice, to add about 35 grams more, making altogether about 60 grams.

On mixing together the recovered alcohol-water mixture, the second fraction and the additional benzene we should have a liquid of approximately the following composition.

$$\begin{array}{lr}
\text{Alcohol} . & 50\cdot8 \\
\text{Benzene} & 60\cdot3 \\
\text{Water} & 4\cdot8 \\
\hline
& 115\cdot9 \\
\end{array}$$

The mixture should now be distilled, the first two fractions collected as before, and the distillation stopped when the second middle point is reached; fraction 3 from the first distillation should then be added to the residue in the still and the distillation continued. The results would, theoretically, be those given in Table 73.

TABLE 73.

	Temperature range.	Weight.	Alcohol.	Benzene.	Water.
			Theoretical composition.		
1.	67·3—70·6° . . .	59·3	12·7	41·8	4·8
2.	70·6—78·2 . . .	29·2	10·7	18·5	
3.	78·2—82·55 . . .	18·9	46·3		
	Residue of alcohol	27·4			

Amount of Alcohol Recovered.—No doubt the quantity of alcohol actually recovered would be somewhat less than this, but it should be at least 20 grams, giving a total of, say, 54 grams out of 103·7. Moreover, the remainder of the alcohol, except the small amount actually lost by evaporation, could be recovered in the form of the alcohol-water mixture of constant boiling point.

Advantage of Larger Quantities.—Working with larger quantities the result would be much better, for fraction 3 remains nearly the same whatever the quantity distilled and fraction 2 need not be made much larger.

Thus, on distilling 300 grams of the alcohol-water mixture, containing 264·7 grams of alcohol, with 340 grams of benzene, the weights of the fractions would be 435·6, 52·1 and 21 respectively; leaving 131·3 grams of pure alcohol, or nearly half the total quantity. Moreover, the second distillation would give a much more satisfactory result. Recovering the alcohol as the binary mixture with water, adding fraction 2 and 95 grams of benzene, and distilling in the same manner as with the smaller quantity, the weights of the fractions would be 153·1, 31·7 and 21·0 respectively, with a residue of 68 grams of alcohol. The first fraction would also be large enough for a third operation which would yield, say, an additional 20 grams of pure alcohol. It would, in fact, be possible to recover about 215 out of the 264·7 grams of alcohol in the pure state, with very little actual loss.

Alcohols—Water—Benzene.—It has been pointed out that the monhydric aliphatic alcohols may be regarded, on the one hand, as alkyl derivatives of water and, on the other, as hydroxyl derivatives of the paraffins; and that, as the complexity of the alkyl group increases, the properties of the alcohols recede from those of water and approach those of the paraffins or of benzene.

Methyl Alcohol.—Methyl alcohol and water may be separated from each other without difficulty by distillation with an efficient still-head, because their properties are so similar, and their boiling points so far apart that a mixture of minimum boiling point is not formed, and, indeed, the boiling point-molecular composition curve is nowhere nearly horizontal.

Methyl alcohol and benzene cannot be separated from each other by distillation, because their properties are so dissimilar that a mixture of minimum boiling point is formed, and, as its boiling point is much lower than that of the binary benzene-water mixture, no separation can be effected by adding water and distilling. A ternary mixture does not come over, but the first fraction still consists of the benzene-alcohol mixture of constant boiling point.

On the other hand the methyl alcohol may readily be extracted from its mixture with benzene by shaking with water, because, although the alcohol is miscible in all proportions both with benzene and with water, it resembles water much more closely, and its solubility in that substance may perhaps be said to be greater than in benzene.

The greater the molecular weight of an alcohol, or, in the case of isomers, the higher the boiling point, the more

difficult is the extraction of the alcohol from a solution in
benzene by means of water; with isobutyl alcohol the
process is very slow indeed.

Ethyl, Isopropyl, Normal Propyl and Tertiary Butyl Alcohol.
—Ethyl, isopropyl, normal propyl and tertiary butyl alcohol
all form binary mixtures of minimum boiling point both with
water and with benzene. Pure ethyl alcohol cannot be
obtained, even from its very strong aqueous solution, by
distillation because its boiling point is so very little higher
than that of the alcohol-water mixture; and, on account of
the similarity of the properties of the two substances,
dehydrating agents act in a very similar manner on them, and
it is only under very special conditions and, apparently, only
with one dehydrating agent—freshly ignited lime—that the
last traces of water can be removed.

Each of the four alcohols, however, forms a ternary
mixture of minimum boiling point with benzene and water,
and the latter substance may be removed from the strong
alcohols by distillation with benzene.

Isobutyl Alcohol.—Isobutyl alcohol, however, does not
form a ternary mixture of ˙minimum boiling point with
benzene and water, and when a mixture of these three
liquids is distilled, the first portion of the distillate consists,
as has been stated, of the benzene-water mixture which
boils at 69·25°.

Higher Alcohols.—Benzene may, in fact, be removed from
isobutyl alcohol or any other of higher boiling point by
adding sufficient water to form the binary benzene-water
mixture, and distilling with an efficient still-head.

General Statement.—Thus, it is possible to extract water
from the lower alcohols—except methyl alcohol—by adding
benzene and distilling, and to extract benzene from the
higher alcohols by adding water and distilling.

It is probable that other cases will be discovered in which one component of a mixture may be removed by distillation with an additional substance.

REFERENCES.

Young, "The Preparation of Absolute Alcohol from Strong Spirit," *Trans. Chem. Soc.*, 1902, **81**, 707.

Young and Fortey, "The Properties of Mixtures of the lower Alcohols with Benzene and with Benzene and Water," *ibid.*, 1902, **81**, 717.

CHAPTER XIX

GENERAL REMARKS

PURPOSES FOR WHICH FRACTIONAL DISTILLATION IS RE-
QUIRED. INTERPRETATION OF EXPERIMENTAL RESULTS.
CHOICE OF STILL-HEAD, NUMBER OF FRACTIONS, &c.

Purposes for which Fractional Distillation is required.—Fractional distillation may be employed for various purposes, of which the following are among the more important.

1. The isolation of a single substance from a mixture with the smallest possible loss.

2. The separation of the components from a mixture of known qualitative composition.

3. The determination of the quantitative composition of a mixture of which the qualitative composition is already known, or the determination of the quantitative composition of a mixture of constant boiling point.

4. A general study of the qualitative and quantitative composition of a complex mixture when only the general nature of the chief components is known, or when it is definitely known that certain substances are present but it is not known what others there may be.

THE SEPARATION OF A SINGLE SUBSTANCE FROM A MIXTURE
WITH THE SMALLEST POSSIBLE LOSS

1. When a pure substance is to be isolated from a mixture, the procedure will be somewhat simplified if the true boiling point of the chief component is already known.

In any case a careful record should be made of the temperature range and weight of each fraction, and we may then either calculate the values of $\Delta w/\Delta t$ (p. 122) or plot the total weights of distillate against the final temperatures of the fractions.

Much time will be saved if an improved still-head is employed; for liquids of high boiling point, the "pear," and for volatile liquids the "evaporator" apparatus may be especially recommended.

Interpretation of Results. First Case.—If the results obtained by the distillation of the liquid under examination give a curve similar in form to Fig. 65, we may conclude that there is present a relatively small quantity of some impurity of much higher boiling point, but that impurities more volatile than the chief component are absent.

As the temperature shows no appreciable rise until towards the end of the distillation, the first fraction will have a perfectly constant boiling point when redistilled, and we may conclude that it will most probably consist of the pure substance required. In the second and subsequent fractionations the greater part of the distillate from the first fraction may, in each case, be taken as most probably pure and will not require to be re-

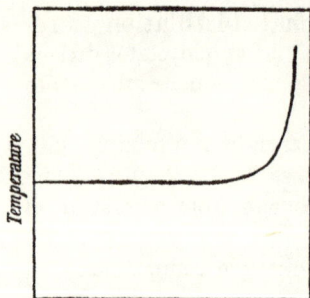

Weight of Distillate.
FIG. 65.

distilled. The separation of the liquid of constant boiling point will in such a case as this be an easy one.

Proofs of Purity.—Even, however, if the true boiling point of the pure substance is known, and if the observed boiling point agrees with it, we cannot be absolutely sure of the purity of the liquid unless we know—from the method of preparation—that there is no possibility of the formation of a mixture which distils without change of composition at practically the same temperature as the pure liquid. If we can be sure that the original mixture contains only substances which are very closely related to each other, such as members of a homologous series, the problem will be much simplified, for it may then be concluded with certainty that no mixture of constant boiling point can be formed.

If any doubt exists and if the specific gravity of the pure liquid is accurately known, it is best to determine that of the distillate : if both the boiling point and the specific gravity of the distillate agree with those of the pure substance, we may conclude with confidence that the distillate is really pure. Instead of the specific gravity, the refractive index, the vapour density, the melting point (if the liquid can be easily solidified), or some other physical constant may be determined for comparison, or a chemical analysis of the liquid may be made.

Hexane and Benzene.—Suppose, for example, that the original liquid consisted chiefly of normal hexane, and that we were unacquainted with the previous history of the specimen ; there would be the possibility of the presence of some benzene. But a mixture of normal hexane and benzene containing as much as 10 per cent. of the aromatic hydrocarbon boils at almost exactly the same temperature as pure hexane, and therefore the presence of the benzene could not be detected, and the impurity could not be removed, by fractional distillation. The high specific gravity of the distillate

would, however, show that some other substance besides
normal hexane was present.

If, on the other hand, it were known that the hexane had
been prepared synthetically from pure propyl iodide by the
action of sodium, there would be no possibility of the presence
of benzene.

Ethyl Alcohol and Water.—Ethyl alcohol containing, say,
15 or 20 per cent. of water would behave in the manner
indicated by the same curve (Fig. 65), but, in this case, even
with the most efficient still-head, we should not have pure
alcohol in the first part of the distillate, nor even the pure
mixture of constant boiling point, but a mixture containing
at least 5 and probably as much as 7 or 8 per cent. of water,
for the mixture of constant boiling point which contains
95·57 per cent. of alcohol is extremely difficult to separate
from water, although there is a wide difference between the
two boiling points.

In this case, again, the boiling point of the binary alcohol-
water mixture is so slightly lower than that of pure alcohol
that the reading of the thermometer—unless very accurate
—would hardly be sufficient to distinguish with certainty
between the two.

Isopropyl Alcohol and Water.—With isopropyl alcohol
and water, similar results would be obtained except that the
difference between the boiling points of the pure alcohol and
of the binary mixture is sufficiently great for the two to be
distinguished without difficulty by the observed temperature,
and also that the binary mixture can be separated in a pure
state from water.

Second Case.—Figure 66 represents the separation of a
liquid of constant boiling point from a mixture which
contains only much more volatile impurities. The separation
is an easier one than the last since it is the least volatile

component that is to be isolated. If the temperature remains constant for a considerable time and shows no rise whatever at the end of the distillation, the last portion need not be redistilled, and in subsequent fractionations the distillation may be stopped as soon as the maximum temperature has been reached and the residue may be taken as pure.

Here, again, it is possible that we may be dealing with a mixture of maximum boiling point as, for example, in

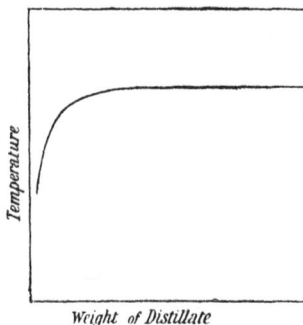

FIG. 66.

the separation of the binary mixture of chloroform and methyl acetate from a slight excess of the ester, but it is less probable, for mixtures of maximum boiling point are not often met with. It is also just possible, though unlikely, that we might have a mixture of minimum boiling point containing neither substance in excess but contaminated with a more volatile impurity.

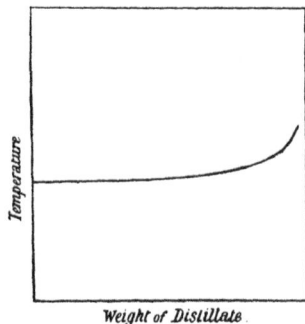

FIG 67.

Third Case.—A curve like that in Fig. 67, would result from the distillation of a liquid containing an impurity, the boiling point of which was not much higher than that of the chief component.

The separation in such a case would be much more difficult, and several fractionations would be required before the boiling point of the first fraction became quite constant. When the boiling points of the components of a mixture are

R 2

very near together and the chemical relationship is not very close, mixtures of constant boiling point are not unlikely to be formed ; that is the case, for example, with carbon tetra-chloride and benzene, and it would be practically impossible to separate either pure tetrachloride or the mixture of constant boiling point from any mixture of the two substances, although benzene, if present in large excess in the original mixture, could be separated in a pure state by repeated fractionation.

Fourth Case.—A curve such as that in Fig. 68, would represent the behaviour on distillation of any mixture—such as

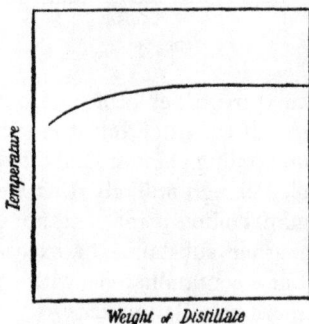

Weight of Distillate

Fig. 68.

that last-named with benzene in large excess—in which the chief component was the less volatile, but in which the difference between the boiling points of the components was small. The separation of the chief component would almost invariably be easier than in cases represented by Fig. 67. Indeed, Fig. 68 bears the same relationship to Fig. 67 as Fig. 66 to Fig. 65.

Here, again, the highest fraction might consist of a mixture of maximum boiling point such as chloroform-methyl acetate ; (b. p. 64·5°) with chloroform (b. p. 60·5°) in excess ; or, less probably, of a mixture of minimum boiling point containing a more volatile impurity, as, for example, the isopropyl alcohol-water mixture (b. p. 80·35°) with a little ethyl alcohol (b. p. 78·3°).

Fifth and Sixth Cases.—The curves in Figs. 69 and 70 represent the distillation of liquids which contain impurities of both greater and less volatility. In the case of Fig 69, the

boiling points of these impurities are far removed from that
of the substance to be separated; in the case of Fig. 70
they are near it.   Such separations as these have very
frequently to be carried out.

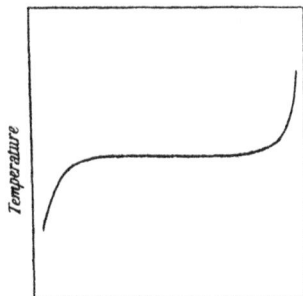

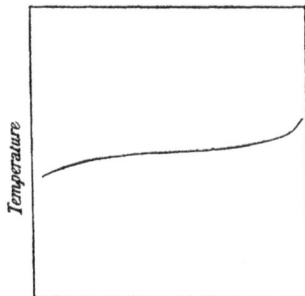

<table>
<tr><td>Weight of Distillate.
Fig. 69.</td><td>Weight of Distillate
Fig. 70.</td></tr>
</table>

When, for example, benzene is nitrated by treatment with
nitric and sulphuric acids, some of the benzene is usually
unacted on, while a certain amount of dinitrobenzene is
formed.   Here the boiling point of the chief component,
nitrobenzene, is very much higher than that of benzene
and far below that of dinitrobenzene, and the separation is
therefore an easy one.   If pure benzene is used for the
preparation and the mixture is distilled through a "pear"
still-head with, say, 12 bulbs, the collection of pure nitrobenzene may be commenced after the second or third
fractionation.

When, however, the boiling points are not so far apart, as,
for example, in the separation of ethyl acetate from methyl
and propyl acetate, the process is tedious if an ordinary
apparatus is used (p. 128), and even with a 5-column
"evaporator" still-head several fractionations are required
(p. 193), for the middle substance is always more difficult to
separate than the others.

If the boiling points of the components are very near

together (Fig. 70) the separation of the middle substance is extremely troublesome, and there is greater probability that mixtures of constant boiling point may be formed than in the cases previously considered.

Other Cases.—Other cases than those referred to may be met with, for example a liquid may contain only impurities of greater volatility but the boiling points of some of these may be near, those of others far below that of the chief component. The most volatile impurities in such a case will be easy to remove, while the less volatile will only be eliminated with difficulty and the distillation will correspond to those represented by the curve in Fig. 68 rather than in Fig. 66.

THE SEPARATION OF THE COMPONENTS OF A MIXTURE OF KNOWN QUALITATIVE COMPOSITION

Two Components.—The chief points to be considered are (a) the boiling points of the pure components, and the difference between them; (b) the chemical relationship of the components; (c) the form of the boiling point-molecular composition curve, if it can be ascertained.

Closely related Substances.—By far the simplest cases are those in which the substances present are chemically closely related to each other, for the form of the boiling point-molecular composition curve must then be normal or nearly so (Chap. IV); there is no possibility of the formation of mixtures of constant boiling point, and the only points to be considered are the actual boiling points of the components and the difference between them.

Boiling Points of Components.—On the actual boiling points will depend the nature of the still-head that can be used; the greater the difference between them, the more easily can the separation be effected.

Substances not Closely Related; Mixtures of Constant Boiling Point.—If the substances are not closely related to each other, mixtures of constant boiling point may be formed and the list of such mixtures which are at present known (p. 67) may be consulted. It will be sufficient to remark here that water forms mixtures of minimum boiling point with the majority of organic compounds ; that, generally, such mixtures are most frequently met with when the molecular weight of one of the components is abnormal in the liquid state, and that this is usually the case with compounds which contain a hydroxyl group.

It should also be remembered that when the boiling points of any two substances are near together, a small deviation from the normal boiling point-molecular composition curve is sufficient to give rise to the formation of a mixture of constant boiling point, or, at any rate, to make the curve nearly horizontal over a large part of its course starting from one extremity, usually that corresponding to the lowest temperature. It has been pointed out (p. 220) that when the curve is of this form it is very difficult, and may be impossible, to separate that component which is in excess where the curve is nearly horizontal.

When a mixture which distils without change of composition is formed, and its boiling point is far below that of the more volatile component, as in the case of normal propyl alcohol and water, or of methyl alcohol and benzene, or far above that of the less volatile component as with nitric acid and water, it is usually possible to separate in a pure state both the mixture of constant boiling point and that component which is in excess. But, if the boiling point of the mixture is very near that of one of the two components, as in the case of ethyl alcohol and water (Fig. 71), or of normal hexane and benzene, it is practically impossible to separate that component in a pure state, and it may be impossible to separate the mixture of constant boiling point even when the component which boils at a widely different temperature is in excess. Thus, in the cases referred to, whatever the composition of the original mixture, it is impossible to obtain normal hexane, ethyl alcohol, the binary

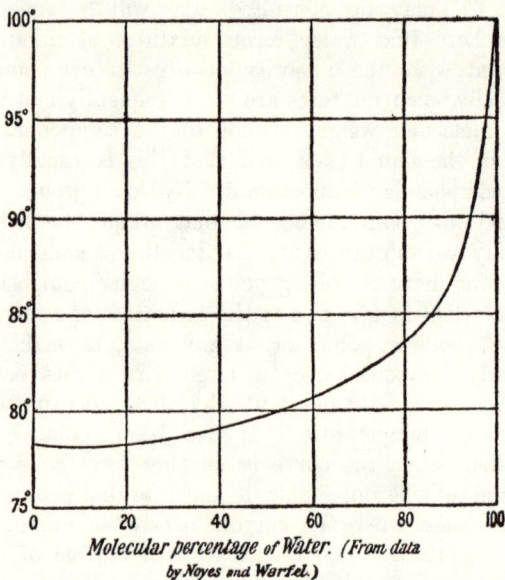

FIG. 71.—Boiling points of mixtures of ethyl alcohol and water.

hexane-benzene mixture or the alcohol-water mixture in a pure state by fractional distillation and it is only the less volatile component, benzene or water, which can be so obtained.

Separation of Components.—The manner in which the composition of the distillate is related to the total amount collected has been discussed in Chap. VIII, and full details of the separation of benzene and toluene by fractional distillation both with an ordinary and an improved still-head have been given (pp. 117 and 191).

Number of Fractions Required and Choice of Still-Head.

—It is difficult to formulate any definite rules regarding the number of fractions in which the distillate should be collected, so many points have to be taken into consideration. The most important of these are the following :—

(a) The efficiency of the still-head employed ;

(b) The approximate quantity of each component ;

(c) The difference between the boiling points of the components ;

(d) The form of the boiling point-molecular composition curve.

Efficiency of Still-Head.—It may be stated quite generally that, for a given mixture, the more efficient the still-head the smaller is the number of fractions that will be required.

Amount of each Component Present.—The approximate quantity of each component must be taken into consideration not only in deciding upon the number of fractions but also in choosing the still-head to be used.

Suppose, for example, that it was desired to separate the benzene as completely as possible from a mixture consisting of 30 grams of that hydrocarbon with 270 grams of toluene.

The best plan would be to employ for the first distillation the most efficient still-head available, and if the recovery of the toluene need not be regarded as of special importance, the complete return of the liquid from the still-head to the still at the end of the distillation would not be essential, and a dephlegmator of many sections might be used. A very slow distillation would be advantageous. It would probably be best to collect all the distillate that came over below 100·0° in one fraction and that from 110·0° to 110·6° in a second. The large residue would consist of pure toluene. The same dephlegmator might be used for the second distillation and the distillate might be collected in three or four fractions, the temperature ranges of which would depend on the efficiency of the still-head. With an 18 section Young and Thomas dephlegmator it is probable that nearly 20 grams of distillate would come over below 81·2°. The fractions would now be small, and it would be necessary to employ a still-head of fewer sections for the remaining distillations. It would be important that the amount of condensed liquid in the still-head should be as small as possible and an evaporator still-head of 5 sections would probably be the most convenient.

If the original mixture contained 270 grams of benzene and 30 grams of toluene, and it was desired to recover the toluene, it would again be best to employ a very efficient still-head, and as, in this case, a most essential point would be that the liquid should return very completely from the still-head to the still at the end of the distillation, an evaporator still-head of many sections would be the best for the purpose.

A mixture of this composition, distilled through a very efficient still-head, would give a considerable amount of nearly pure benzene; so far as the recovery of toluene is concerned, the first 100 grams, or even more, would not be worth redistilling. After rejecting this, fractions might be collected below 80·5°, from 80·5° to 95·6°, and above 95·6°. If the temperature reached 110·6° before the end of the distillation the residue would consist of pure toluene; if not, it would require to be redistilled. In the second fractionation, the distillate from the first fraction of I. would consist of nearly pure benzene and might be rejected. The remaining fractions would now be small and an evaporator still-head of not more than 5 sections would have to be used for the remaining distillations.

Boiling Points of Components.—For mixtures of closely related substances, or others which behave normally on distillation, the greater the difference between the boiling points the more readily can the components be separated by distillation. Thus, it has been shown (p. 153) that a mixture of normal and isopentane (b. p. 36·3° and 27·95° respectively) would require a much larger number of fractions than one of benzene and toluene if a still-head of the same efficiency were employed for both.

Boiling Point-Composition Curve.—When the form of the boiling point-molecular composition curve is normal it is steeper in the region of high temperature than of low, and in general the temperature ranges above the middle point may be somewhat greater than below ; in other words, the number of fractions may be somewhat smaller. When the actual boiling points of mixtures of the two substances are lower than those given by the normal curve, this difference above and below the middle point becomes accentuated ; and when the curve is very flat near its lower extremity, the number of fractions in the low temperature region must be considerably increased while those above the middle point may be diminished in number. In most cases, however, data for the construction of the boiling point-molecular composition curve are wanting, and, unless, we can judge from the nature of the substances in the mixture whether the deviation of the actual from the normal curve is likely to be large or not, we cannot decide on the number of fractions that may ultimately be required until after the fractionation has made some progress.

Three Components.—If there are three substances present and the mixture behaves normally on distillation, the least volatile component will be the easiest to separate and the substance of intermediate boiling point the most difficult.

If, however, the components are not closely related to each other, one or more mixtures of constant boiling point may be formed and the problem becomes more complicated.

Thus, when a mixture of ethyl alcohol, benzene and water is distilled it tends to separate into (a) the ternary mixture of constant boiling point, (b) one of the three possible binary mixtures of constant boiling point, (c) that pure component which is in excess. It may, however, happen that the quantities ʹin the original mixture are such that we have only the two fractions a and b, or a and c, or the fraction a alone. There are, in fact, 12 different ways in which separation may take place, or the mixture may distil unchanged (p. 217).

Again, the substances present may be capable of forming one or two binary mixtures but no ternary mixture of constant boiling point. That would be the case, for example, with isoamyl alcohol, benzene and water, for the only mixtures of constant boiling point that can be formed are those of benzene and water (b. p. 69·25°) or of water and amyl alcohol (b. p. 95·15°). There are, therefore five different ways in which separation may occur, but under no conditions can the mixture distil without change of composition. Employing the initial letters A, B and W for the components —alcohol, benzene and water—we have the following possible separations.

First fraction.	Second fraction.	Third fraction.
$B.W.$	$A.W.$	$A.$
$B.W.$	$A.W.$	$W.$
$B.W.$	$B.$	$A.$
$B.W.$	$A.W.$	...
$B.W.$	$A.$	...

If only a single binary mixture of constant boiling point can be formed, as would be the case, for example, with a mixture of ethyl alcohol (A_1), benzene, and isoamyl alcohol (A_2), the following three separations would be possible.

First fraction.	Second fraction.	Third fraction.
A_1B	A_1	A_2
A_1B	B	A_2
A_1B	A_2	...

A detailed description of the separation of the three

closely related liquids, methyl acetate, ethyl acetate and propyl acetate (both with an ordinary and an improved still-head) is given in Chapters VII and XIV.

3. The use of FRACTIONAL DISTILLATION as a METHOD OF QUANTITATIVE ANALYSIS has been fully discussed in Chapter XVI.

4. A GENERAL STUDY OF THE QUALITATIVE AND QUANTITATIVE COMPOSITION OF A COMPLEX MIXTURE WHEN ONLY THE GENERAL NATURE OF THE CHIEF COMPONENTS IS KNOWN, OR WHEN IT IS DEFINITELY KNOWN THAT CERTAIN SUBSTANCES ARE PRESENT, BUT IT IS NOT KNOWN WHAT OTHERS THERE MAY BE.

Rough Estimate of Composition.—When a complex mixture, the qualitative composition of which is only partly known, is distilled through an efficient still-head and the distillate is collected in fractions of either equal temperature range or of approximately equal weight, a rough estimate of the boiling points and of the amounts of the components may frequently be obtained from the quantities of distillate in the first case, or from the temperature ranges in the second, or, generally, from the values of $\Delta w/\Delta t$.

Causes of Confusion.—But confusion is apt to arise when (*a*) there are two substances present with boiling points very near together, (*b*) one of the components is present in relatively very small amount, or (*c*) two or more of the substances form mixtures of constant boiling point.

a. **Two Components Boiling at nearly the same Temperature.**—*If there are isomeric members of a homologous series present in the mixture, it is very likely that their boiling points may in some cases be very near together.*

Pentanes in Petroleum.—An instance of this has already been referred to (1) in the case of the light distillate from

American petroleum consisting chiefly of butanes, pentanes and hexanes. Even with the very efficient still-head shown in Fig. 45 (p. 183), the results of the first distillation seem to indicate the presence of only a single substance between the butanes and hexanes, boiling at about 33°; and it is only after repeated fractionation that the presence of both normal and isopentane (boiling points 36·3° and 27·95°) is clearly shown. The fact, however, that for the middle fractions the values of $\Delta w/\Delta t$ diminish, while for those below and above them they increase, is a clear indication that we are not dealing with a single substance.

This may be seen from Table 74, in which the results of the first three fractionations are given; it will be seen that, in the first fractionation, the highest values of $\Delta w/\Delta t$ are those for fractions 5 and 6; in the second, the fractions that have the highest values are Nos. 5 and 7, and in the third they are Nos. 4 and 7.

TABLE 74.

	I.			II.		III.	
Number of fraction.	Final tempe- rature = t	$\dfrac{\Delta w}{\Delta t}$		t	$\dfrac{\Delta w}{\Delta t}$	t	$\dfrac{\Delta w}{\Delta t}$
1	...	...		28·05°	?	27·95°	?
2	28·5°	?		29·15	68	29·15	35
3	29·9	72		30·55	71	30·5	74
4	31·3	58		31·7	123	31·5	172
5	32·85	179		32·45	204	32·3	135
6	33·85	242		33·5	140	33·3	110
7	...	...		34·45	168	34·05	193
8	35·25	126		35·4	142	34·9	142
9	36·8	74		36·8	81	36·1	117
10	41·1	17		37·8	54	37 1	88

Considering 5 and 6 as one fraction the values of $\Delta w/\Delta t$ for the first three fractionations would be 204, 167, 121, respectively; the values for No. 4 in the three fractionations are 58, 123 and 172, and for No. 7 they are 168 and 193 in the second and third fractionations.

The whole course of the separation is, however, better seen in the diagram (Fig. 29, p. 139). Both the pentanes may be obtained in a pure state by distillation only, or, at any rate, after removal of small quantities of impurity by means of a mixture of nitric and sulphuric acids. It does not appear, however, that any hydrocarbon that boils at a higher temperature than normal pentane can be so separated.

Hexanes in Petroleum.—Let us consider, for example, the hexanes (2). A preliminary distillation through an ordinary or fairly efficient still-head appears to indicate the presence of a single substance boiling at about 66°, and after a few fractionations with a very efficient still-head it is seen that a further separation, apparently into two components, as in the case of the pentanes, is taking place, but after long continued fractionation it is found that the separation does not end there. Two series of fractionations were carried out, one of American the other of Galician petroleum.

Material Employed.—The American petroleum at first came over for the most part between 28° and 95°, and was richest in hexanes; the weight of distillate between 56° and 74° was about 800 grams. Aromatic hydrocarbons were not removed before the fractionation by treatment with nitric and sulphuric acids.

From the Galician petroleum, the hydrocarbons boiling below 40° and above 72° had previously been for the most part separated by distillation, and benzene was removed before the fractionation was commenced. The weight of the Galician petroleum was about 5 times as great as that of the American.

Still-heads Used.—The fractionation of the American petroleum was carried out with a 12-section Young and Thomas dephlegmator, that of the Galician with the combined dephlegmator and regulated temperature still-head which was employed for the separation of the pentanes (Fig. 45). The Galician petroleum was thus freer from pentanes and heptanes to begin with than the American, and the still-head employed was more efficient.

Description of Results.—In Table 75 are given the temperature ranges and the values of $\Delta w/\Delta t$ for the fourth fractionation of the American petroleum, and it will be seen that at this early stage the fraction from 65° to 66° has the highest value, but in the subsequent fractionations the ratio for the corresponding fraction (above 65°) steadily fell, and after the tenth fractionation this fraction had the lowest value.

TABLE 75.

IV.		IV.	
Temperature range.	$\dfrac{\Delta w}{\Delta t}$	Temperature range.	$\dfrac{\Delta w}{\Delta t}$
57 —60°	9·1	66 —67°	80·5
60 —61·5	31·8	67 —68	87·1
61·5—63	54·1	68 —69	90·2
63 —64	56·1	69 —70·5	60·7
64 —65	94·7	70·5—73	24·3
65 —66	102·0		

Graphical Representation.—So far as the middle fraction is concerned, the gradual change in the value of $\Delta w/\Delta t$ closely resembles that observed in the case of the two pentanes, but in other respects the fractionations show marked differences, which become apparent when the curves (Fig. 72) representing the separation of the hexanes are compared with those for the pentanes (Fig. 29, p. 139).

As the quantity of Galician petroleum was much greater than that of American, the weight of distillate coming over between 60° and 70° has been taken as 100 for both of them in each fractionation, and as the tenth and sixteenth frac-

tionations of American petroleum correspond very closely
with the fourth and seventh respectively of the Galician,
only one curve has been drawn for each of these two pairs,
the experimental results for American petroleum being
indicated in these and other cases by crosses, and for Galician
by circles. The fractionations of American petroleum are
referred to as IV_A, X_A . . . and of Galician as IV_G, VII_G . . . .

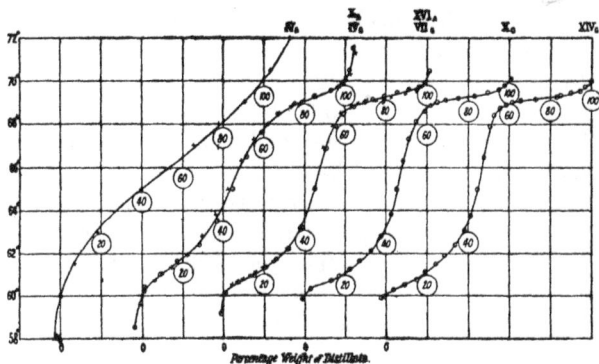

FIG. 72.—Results obtained by fractional distillation of American and of Galician
petroleum. (Fractions from 58° to 72°.)

After the fifth fractionation of the pentanes, the values of
$\Delta w/\Delta t$ for the lowest and highest fractions show a steady
rise, until, in the thirteenth fractionation, the last portion of
liquid boiled quite constantly at 36·3°, and, in the eighteenth,
the first portion of distillate came over quite constantly at
27.95°; in other words, the curves became quite horizontal
at these two temperatures.

Highest Fractions.—With the hexanes, on the other hand,
no fraction was obtained with perfectly constant boiling
point even though, in the case of American petroleum, the
liquid which came over above 66° was fractionated thirty-
one times. Indeed, after the sixteenth fractionation, the
maximum value of $\Delta w/\Delta t$ (at about 69·1°) showed very little

S

further rise, and after the twenty-first fractionation it diminished slightly.

With the Galician petroleum, no further improvement in the value of $\Delta w/\Delta t$ for the fraction a little above 69° was noticeable after the eleventh fractionation.

Lowest Fractions.—Again, at the lower temperatures, in the case of American petroleum, the highest individual value of $\Delta w/\Delta t$ was reached in the twelfth fractionation, (189 for the fraction coming over between 60·85° and 61·0°). In the sixteenth fractionation, the highest value had fallen to 141 for the fraction from 60·75 to 60·9, but, on the other hand, the values for the fractions above and below these temperatures showed a tendency to rise.

With Galician petroleum the highest individual value of $\Delta w/\Delta t$ at the lower temperatures was reached in the eleventh fractionation (790 for the fraction from 60·6° to 60·85°). In the last fractionation, the seventeenth, the maximum was only 626 for the fraction coming over between 60·55° and 60·75°, but in this case also the values for the higher and lower fractions appeared to be increasing slightly.

Interpretation of Results.—These facts may be explained by assuming the presence of at least four substances in both American and Galician petroleum boiling at temperatures between, say, 59° and 72°, two of them with boiling points not far from 61° and two not far from 69·5°. We should then have at first the apparent separation of a single substance boiling at about 66°, just as with the pentanes we have in the first place apparently the separation of a single component boiling at about 33°. Later on there would appear to be a further separation into two components with the approximate boiling points 61° and 69·5° corresponding to the two pentanes which boil at 27·95° and 36·3° respectively.

But while the two pentanes can be separated in a pure state, the two components from the hexane fractions cannot; and, in the later fractionations, we have the beginning of a further subdivision. But each of the final separations must necessarily be far more difficult than the earlier ones, for the boiling points of the components of each pair are much closer together than the apparent boiling points of the pairs themselves, and although a partial separation may be effected, it is extremely doubtful whether, with the most efficient still-head, any one of the components could be separated in a pure state by fractional distillation alone.

Isohexane and Diethyl-methyl Methane.—As regards the components boiling not far from 61°, there is the possibility of the presence of the two isomers with the formulæ $(CH_3)_2CH-CH_2-CH_2-CH_3$ and $(C_2H_5)_2CH-CH_3$, which would certainly resemble each other very closely indeed in their physical properties. It is doubtful, however, whether either of these paraffins has yet been prepared in a perfectly pure state, and sufficient reliance cannot be placed on the accuracy of the determinations of their boiling points and specific gravities, which have been made up to the present time, to warrant the formation of any estimate as to the relative quantities of the two hydrocarbons from the temperature ranges or specific gravities of the fractions.

Di-isopropyl.—It is noticeable, however, that the observed specific gravities (minimum 0·6728 at 0°/4° for the fraction from 60·55° to 60·75°) are slightly higher than the most probable value for isohexane (about 0·6721 at 0°/4°), and this may be due to the presence of a little di-isopropyl, $(CH_3)_2CH-CH(CH_3)_2$, which boils at 58·1° and has a somewhat higher specific gravity (0·67948 at 0°/4°), or it is just possible that the pentamethylene (b. p. about 50·5°, sp. gr. 0·7506 at 20·5°/4°), which is certainly present in petroleum, may not have been completely removed.

Normal Hexane and another Component.—With regard to the fractions boiling near to 69·5°, we have the following facts to guide us. The boiling point of pure normal hexane is 68·95° and its specific gravity at 0°/4° is 0·67697; but in the last fractionation of the Galician petroleum the fraction from 69·12° to 69·20° had the highest value of $\Delta w/\Delta t$, and its specific gravity at 0°/4° was about 0·685.

The less volatile portion of the American petroleum was fractionated thirty-one times; the fractions were then separately treated several times with mixed nitric and sulphuric acids, and were subsequently fractionated many times with the object of separating pure normal hexane. The best specimen that could be obtained boiled at 69·05° and had the specific gravity 0·67813 at 0°/4°.

In both cases the specific gravities of the fractions above 66° were higher than those of pure normal hexane. For the fractions from 66° to 69° the change was slight, but above 69° the specific gravity rose rapidly. The following determinations were made in the course of the fractionations.

TABLE 76.

Temperature range.	Sp. gr. at 0°/4°.	Temperature range.	Sp. gr. at 0°/4°
66·4 —67·85°	0·6793	69·4 —69·55°	0·6898
68·6 —68·85	0·6802	69·5 —69·7	0·6962
68·95—69·03	0·6803	69·7 —69·95	0·7095
69·0 —69·1	0·6815	69·95—70·15	0·7157
69·25—69·4	0·6856	70·2 —74·0	0·7306

The fact that the fractions which showed the highest values of $\Delta w/\Delta t$ came over at temperatures so little higher than the boiling point of pure normal hexane (American petroleum about 69·1°; Galician, about 69·2°) shows that there may be either a considerable quantity of some other substance present with a boiling point only a fraction of a degree higher than that of normal hexane, or there may be a smaller quantity of a substance which boils not more than 2° or 3° higher. That the separation is such a difficult one shows that the difference of boiling point can hardly be greater than three degrees.

Evidence afforded by Specific Gravities.—Again if the specific gravities of the fractions given in Table 76 are plotted against the mean temperatures it will be seen that there is a very sudden rise of specific gravity above 69°, and this seems also to show that the boiling point of the second substance cannot be far from that of normal hexane. If, for example, the substance boiled at a temperature as high as 80° the rise would certainly be more gradual at first. Moreover, it has been shown (p. 140) that hexamethylene, which boils at 80·85°, can be separated from the hexanes by fractional distillation.

Methyl Pentamethylene.—Further light has been thrown on the problem by the researches of Markownikoff (3), of Zelinsky (4), and of Aschan (5) on Russian petroleum, which has been found to contain large quantities of hexamethylene (b. p. 80·85° ; sp. gr. 0·7968 at 0°/4°) and methyl pentamethylene (b. p. about 71·5° ; sp. gr. about 0·766 at 0°/4°), both of which substances have also been prepared synthetically. The two hydrocarbons differ widely in their behaviour with fuming nitric acid, for hexamethylene is only attacked slowly even when heated, a large amount of adipic acid being formed, whereas methyl pentamethylene is acted on rapidly at the ordinary temperature, with evolution of much heat, and acetic acid is the chief product of oxidation.

Now the fractions for the first three or four degrees above 69° are attacked in the cold by fuming nitric acid, much heat being evolved and a large amount of acetic acid formed, and it may therefore be concluded that the substance present in American and Galician petroleum with a boiling point not far above that of normal hexane is methyl pentamethylene.

Preparation of Pure Normal Hexane.—Not only methyl pentamethylene, but also the isohexanes and other hydrocarbons which contain a $>$CH— group may be removed by

heating with fuming nitric acid, and it was found that when
the fractions from Galician petroleum which came over
between 66° and 69·2° were subjected to prolonged heating
with the fuming acid, and were afterwards distilled two or
three times, almost pure normal hexane was obtained;
indeed from the fractions between 66° and 68·95° the normal
paraffin appeared to be perfectly pure.

Summary of Results.—It has thus been shown that in the
distillation of ·this portion of American or Galician
petroleum, the liquid which at first seems to be a single
substance boiling at about 66° proves to be a mixture of
four substances, two isomeric hexanes boiling at nearly the
same temperature, 61°, and normal hexane and methyl
pentamethylene with the boiling·points 68·95° and (about)
71·5° respectively. The first pair of substances are very
closely related to each other, the second pair are not.

Amyl Alcohols in Fusel Oil.—As another example of a
pair of very closely related liquids which are frequently met
with, the isomeric amyl alcohols which are present in fusel
oil may be mentioned. Distilled through an ordinary still-
head, long continued fractionation would be necessary even
to indicate the presence of two isomeric amyl alcohols, and
the boiling points are so close together that even with an
exceedingly efficient still-head the separation is very
difficult.

Hexamethylene and a Volatile Heptane.—Another example
of the presence, in a complex mixture, of two substances
which boil at nearly the same temperature is afforded by
the distillation of the portion of American or Galician
petroleum that comes over between 75° and 80°. It was
found possible to separate a liquid of quite constant boiling
point and as derivatives of hexamethylene could be prepared
from it (6), it was concluded that that substance had been
separated in a pure state. Later on, however, it was found

that the liquid could be partially but not completely frozen
in an ordinary freezing mixture and eventually, by fractional
crystallisation, pure hexamethylene was obtained with
practically the same boiling point as before but with a definite
melting point, and of notably higher specific gravity (7). It
was evident that the substance separated by fractional
distillation only was a mixture of two hydrocarbons, hexamethylene and a heptane, of which the former was present
in much the larger quantity, and that either the boiling
points are almost identical, or else the two substances form
a mixture of constant boiling point almost identical with
that of hexamethylene.

Pentamethylene and Trimethyl-ethyl-methane.—Such
difficulties as have been described are of common occurrence
in the distillation of petroleum. For example, American,
Galician and Russian petroleum all contain a certain amount
of pentamethylene which boils at about 50° ; but there is
also present a hexane, trimethyl-ethyl-methane, boiling at
nearly the same temperature, and it appears to be impossible
to separate these hydrocarbons by fractional distillation.

b. **One or more Components Present in Small
Quantity.**—*If one or more components of a complex mixture
are present in relatively very small amount, they are apt to be
altogether overlooked, and it is only by keeping a careful record
of the weights and temperature ranges of the fractions, and
by calculating the values of $\Delta w/\Delta t$ or mapping the total
weights of distillate against the temperatures, that the existence
of these components can be detected.*

The manner in which the presence of a relatively very
small quantity of hexamethylene was recognised in
American petroleum has been described in Chap. VII
(p. 140).

Pentamethylene in Petroleum.—In a similar manner, it
was found that there is a small quantity of pentamethylene

(and also trimethyl-ethyl-methane) in American petroleum (2).

The first fractionation of all the available distillates from petroleum ether obtained with the combined dephlegmator and regulated temperature still-head, and collected between about 37° and 60°, gave no indication of the presence of any substance boiling in the neighbourhood of 50°, as will be seen by Table 77.

TABLE 77.

I.		II.		IV.		VII.	
Temperature range.	$\dfrac{\Delta w}{\Delta t}$	Temperature range.	$\dfrac{\Delta w}{\Delta t}$	Temperature range.	$\dfrac{\Delta w}{\Delta t}$	Temperature range.	$\dfrac{\Delta w}{\Delta t}$
36 —37°	47·4	36 —37°	47·8	36·8—42·0°	5·0	45·0 —49·35°	3·8
37 —40	22·2	37 —39	11·8	42·0—47·7	3·1	49·35—50·1	34·9
40 —45	3·1	39 —42	5·9	47·7—50·7	12·9	50·1 —51·3	16·0
45 —50	6·2	42 —47	4·0	50·7—53·4	12·6	51·3 —53·7	6·7
50 —54	12·2	47 —51	6·0	53·4—56·2	7·8		
54 —57·5	22·6	51 —54	16·5	56·2—58·6	10·3		
57·5—59·5	24·7	54 —57	14·4	58·6—59·6	15·7		
59·5—60·4	39·0	57 —59·2	17·8				
		59·2—60·1	38·9				

FINAL FRACTIONATION.

Temperature range.	Δw	$\dfrac{\Delta w}{\Delta t}$	Specific gravity at 0°/4°.
47·3 —49·45°	11·5	5·3	0·7029
49·45—49·55	16·3	163·0	0·7035
49·55—50·35	16·0	20·0	0·6975
50·35—56·4	15·0	2·5	

In the second distillation, however, the value of $\Delta w/\Delta t$ for the fraction 51—54° was slightly higher than for the one above and much higher than for that below it.

In the fourth fractionation the maximum was quite clearly defined, and the value of $\Delta w/\Delta t$ near 50° steadily increased in the subsequent operations, rising from 12·9 in the fourth to 34·9 in the seventh and 48·0 in the ninth.

The quantity of material was too small to allow of the separation being continued with the combined dephlegmator and regulated temperature still-head, but six additional fractionations were carried out with a 12 column dephlegmator, when the value of $\Delta w / \Delta t$ rose to 163·0 for the fraction from 49·45° to 49·55°.

The distillate, amounting to 16·3 grams, was found to yield glutaric acid on oxidation with nitric acid, and therefore contained pentamethylene. The specific gravity, 0·7035 at 0°/4° was, however, too low and the vapour density, 39·2, too high for pure pentamethylene (sp. gr. 0·751 at 15°/15° ; vap. den. 35). As the calculated vapour density of hexane is 43, it would appear that the distillate consisted of a mixture of pentamethylene and trimethyl-ethyl-methane in approximately equal molecular proportions. The results are quite in accordance with those obtained by Markownikoff (8), who showed that both these hydrocarbons are present in Russian petroleum, the quantity of pentamethylene and other naphthenes as compared with that of the paraffins, being, however, much larger in Russian than in American petroleum.

c. **Mixtures of Constant Boiling Point.**-–*When two or more of the components form mixtures of constant boiling point the difficulty in interpreting the experimental results is greatly increased.*

Benzene in American Petroleum.—For example, if we subject a quantity of American petroleum to distillation, collecting fractions from 30—60°, 60—90° and 90—120°, and then treat each fraction with a mixture of strong nitric and sulphuric acids, we shall find that a considerable amount of dinitrobenzene will be obtained from the fraction from 60—90° and of dinitrotoluene from that between 90° and

120°. The obvious conclusion from these results is that benzene and toluene are present in American petroleum and this was, in fact, stated long ago to be the case by Schorlemmer.

If, however, instead of collecting the distillate in three large fractions, we were to separate it into nine with a range of 10 degrees each, it would then be found, after repeated distillation, that on treating each fraction with the mixed acids, dinitrobenzene would be obtained almost exclusively from the distillate that came over between 60° and 70° and dinitrotoluene from that between 90° and 100°, although the boiling points of benzene and toluene are 80·2° and 110·6° respectively.

Possible Explanations of Results.—In order to explain these facts two assumptions may be made; either (a) it is not benzene and toluene that are present in American petroleum, but some other more volatile compounds from which dinitrobenzene and dinitrotoluene respectively are formed by the action of nitric and sulphuric acids; or (b) benzene and toluene, when mixed with the paraffins present in petroleum come over chiefly at temperatures 14° or 15° below their true boiling points.

The first assumption, so far as the conversion of a hydrocarbon other than benzene into dinitrobenzene is concerned, was actually made by one observer, but it is practically certain that the second is the correct one, for it has been found, as already stated, that when a mixture of normal hexane (b. p. 68·95°) with, say, 10 per cent. of benzene, is distilled, the benzene cannot be separated but comes over with the hexane at 68·95°. Lastly, the fact that when American petroleum is distilled, the greater part of the benzene comes over below 69° (mostly at 65—66°) is explained by the presence of isomeric hexanes which boil at lower temperatures than normal hexane and also carry down the benzene with them.

Ethyl Alcohol—Benzene—Water.—The behaviour of mixtures of ethyl alcohol, benzene and water has already been referred to (p. 217); but it may be pointed out that, when a mixture of equal weights of benzene and of 95 per cent. alcohol is distilled through a very efficient still-head, the substance of highest boiling point, water, comes over in the first of the three fractions into which the distillate tends to separate, the remainder of the benzene in the second fraction, while the third fraction or residue consists of the most volatile of the original components, alcohol.

Aliphatic Acids and Water.—Again, when a mixture of formic, acetic and butyric acids with water is distilled, it tends to separate into three or more of the following components.

		Boiling point.
1. Butyric acid-water (mixture of minimum boiling point).		99·2°
2. Water		100·0
3. Formic acid		100·7
4. Formic acid-water (mixture of maximum boiling point).		107·1
5. Acetic acid		119·0
6. Butyric acid		163·8

With a large amount of water, the whole of the butyric acid would come over in the lowest fraction, and if the amount of acetic acid was large, the last fraction would consist of that acid; but although acetic acid does not form a mixture of minimum boiling point with water, yet it is very difficult to separate the acid from its dilute aqueous solution by distillation, and if the amount of the acid was relatively small it might all be carried down with the water at temperatures below 107°, and in that case the highest fraction would consist of the formic acid-water mixture of maximum boiling point. The acids would then come over in the reversed order of their boiling points. Hecht (9) found that on distilling a mixture of acetic, butyric and œnanthylic acids with much water, the whole of the œnanthylic acid came over in the first portion of the distillate, the middle portion contained chiefly butyric acid and the last portion contained acetic acid nearly free from the other two. Hecht points out that acetic acid is miscible with water in all proportions with considerable heat evolution ; butyric acid is also miscible with water in all proportions but very little heat change is observable, and œnanthylic acid is nearly insoluble in water.

Alcohols and Water.—All the alcohols—with the exception of methyl—form mixtures of minimum boiling point with water, but as the boiling points of the binary mixtures, up to amyl alcohol at any rate, follow the same order as those of the alcohols themselves, such reversals as are observed with the acids do not occur.

As mixtures of constant boiling point are of frequent occurrence among liquids which are chemically not closely related to each other, the possibility of their formation must always be borne in mind in interpreting the results of a fractional distillation.

REFERENCES.

1. Young and Thomas, "Some Hydrocarbons from American Petroleum, I, Normal and Isopentane," *Trans. Chem. Soc.* 1897, **71**, 440.
2. Young, "Composition of American Petroleum," *Trans. Chem. Soc.*, 1898, **73**, 905.
3. Markownikoff, "On Methyl Cyclo-Pentane from different Sources and some of its Derivatives," *Berl. Berichte*, 1897, **30**, 1222.
4. Zelinsky, "Researches in the Hexamethylene Group," *Berl. Berichte*, 1897, **30**, 387.
5. Aschan, "On the Presence of Methyl Pentamethylene in Caucasian Petroleum Ether," *Berl. Berichte*, 1898, **31**, 1803.
6. Fortey, "Hexamethylene from American and Galician Petroleum," *Trans. Chem. Soc.*, 1898, **73**, 103.
7. Young and Fortey, "The Vapour Pressures, Specific Volumes and Critical Constants of Hexamethylene," *Trans. Chem. Soc.*, 1899, **75**, 873.
8. Markownikoff, "On Cyclic Compounds," *Liebig's Annalen*, 1898, **301**, 154.
9. Hecht, "On Isoheptoic Acid from β-Hexyl Iodide," *Liebig's Annalen*, 1881, **209**, 321.

APPENDIX

Correction of Height of Barometer for Temperature.—The atmospheric pressure is usually expressed in millimetres of mercury read at, or corrected to, 0°. In practice the height, H, of the barometer is read at the temperature of the room, t, and the value at 0° calculated by means of one of the two following equations:—

1. $H_0 = H - 0{\cdot}000172\,Ht$ if the scale is etched on the glass; or

2. $H_0 = H - 0{\cdot}000161\,Ht$ if a brass scale is used.

In Table 78 the values of $0{\cdot}000172\,Ht$ are given for each degree from 10° to 30° and for pressures at intervals of 10 mm. from 720 to 780 mm.; and in Table 79 the values of $0{\cdot}000161\,Ht$ for the same temperatures and pressures. If the readings are smaller than 720 mm., it is convenient to plot the corrections against the pressures from 0 to, say, 800 mm. for alternate degrees from 10° to 30° on curve paper. The correction for any pressnre up to 800 mm., at any temperature between 10° and 30° may then be easily ascertained from the diagram.

TABLE 78.

GLASS SCALE. —VALUES OF $0.000172\ Ht$.

t	Height of column of mercury.						
	720	730	740	750	760	770	780
10	1·25	1·25	1·25	1·3	1·3	1·35	1·35
11	1·35	1·4	1·4	1·4	1·45	1·45	1·45
12	1·5	1·5	1·55	1·55	1·55	1·6	1·6
13	1·6	1·65	1·65	1·7	1·7	1·7	1·75
14	1·75	1·75	1·8	1·8	1·85	1·85	1·9
15	1·85	1·9	1·9	1·95	1·95	2·0	2·0
16	2·0	2·0	2·05	2·05	2·1	2·1	2·15
17	2·1	2·15	2·15	2·2	2·2	2·25	2·3
18	2·25	2·25	2·3	2·3	2·35	2·4	2·4
19	2·35	2·4	2·4	2·45	2·5	2·5	2·55
20	2·5	2·5	2·55	2·55	2·6	2·65	2·7
21	2·6	2·65	2·65	2·7	2·75	2·75	2·8
22	2·7	2·75	2·8	2·85	2·9	2·9	2·95
23	2·85	2·9	2·95	2·95	3·0	3·05	3·1
24	2·95	3·0	3·05	3·1	3·15	3·2	3·2
25	3·1	3·15	3·2	3·2	3·25	3·3	3·35
26	3·2	3·25	3·3	3·35	3·4	3·45	3·5
27	3·35	3·4	3·45	3·5	3·55	3·6	3·65
28	3·45	3·5	3·55	3·6	3·65	3·7	3·75
29	3·6	3·65	3·7	3·75	3·8	3·85	3·9
30	3·7	3·75	3·8	3·85	3·9	3·95	4·0

TABLE 79.

BRASS SCALE.—VALUES OF 0·000161 Ht.

t	Height of column of mercury.						
	720	730	740	750	760	770	780
10	1·15	1·2	1·2	1·2	1·2	1·25	1·25
11	1·3	1·3	1·3	1·35	1·35	1·35	1·4
12	1·4	1·4	1·45	1·45	1·45	1·5	1·5
13	1·5	1·55	1·55	1·55	1·6	1·6	1·65
14	1·6	1·65	1·65	1·7	1·7	1·75	1·75
15	1·75	1·75	1·8	1·8	1·85	1·85	1·9
16	1·85	1·9	1·9·	1·95	1·95	2·0	2·0
17	1·95	2·0	2·05	2·05	2·1	2·1	2·15
18	2·1	2·1	2·15	2·15	2·2	2·25	2·25
19	2·2	2·25	2·25	2·3	2·3	2·35	2·4
20	2·3	2·35	2·4	2·4	2·45	2·5	2·5
21	2·45	2·45	2·5	2·55	2·55	2·6	2·65
22	2·55	2·6	2·6	2·65	2·7	2·75	2·75
23	2·65	2·7	2·75	2·75	2·8	2·85	2·9
24	2·8	2·8	2·85	2·9	2·95	3·0	3·0
25	2·9	2·95	3·0	3·0	3·05	3·1	3·15
26	3·0	3·05	3·1	3·15	3·2	3·2	3·25
27	3·15	3·15	3·2	3·25	3·3	3·35	3·4
28	3·25	3·29	3·35	3·4	3·45	3·45	3·5
29	3·35	3·4	3·45	3·5	3·55	3·6	3·65
30	3·5	3·5	3·55	3·6	3·65	3·7	3·75

INDEX

INDEX

When there are several references, the more important are indicated by the numbers being printed in thick type. Numbers printed in italics refer to the bibliographies at the ends of the chapters. Names of authors are printed in small capitals.

T 2

www.ingramcontent.com/pod-product-compliance
Lightning Source LLC
Chambersburg PA
CBHW021425180326
41458CB00001B/142